THE CITY & GUILDS TEXTBOOK

LEVEL 3 NVQ DIPLOMA IN
ELECTROTECHNICAL TECHNOLOGY 2357
UNITS 305–306

About City & Guilds

City & Guilds is the UK's leading provider of vocational qualifications, offering over 500 awards across a wide range of industries, and progressing from entry level to the highest levels of professional achievement. With over 8500 centres in 100 countries, City & Guilds is recognised by employers worldwide for providing qualifications that offer proof of the skills they need to get the job done.

Equal opportunities

City & Guilds fully supports the principle of equal opportunities and we are committed to satisfying this in all our activities and published material. A copy of our equal opportunities policy statement is available on the City & Guilds website.

First edition 2015

ISBN 978-0-85193-279-8

Publisher: Charlie Evans

Development Editor: Hillary Tolan

Production Editor: Fiona Freel

Picture Research: Hillary Tolan

Project Management and Editorial series team: Anna Clark, Joan Miller and Vicky Butt

Cover design by Select Typesetters Ltd

Text design by Design Deluxe, Bath

Indexed by Indexing Specialists (UK) Ltd

Illustrations by Saxon Graphics Ltd and Ann Paganuzzi

Typeset by Saxon Graphics Ltd, Derby

Printed in the UK by Cambrian Printers Ltd

British Library Cataloguing in Publication Data

A catalogue record is available from the British Library.

Publications

For information about or to order City & Guilds support materials, contact 0844 534 0000 or centresupport@cityandguilds.com. You can find more information about the materials we have available at www.cityandguilds.com/publications.

Every effort has been made to ensure that the information contained in this publication is true and correct at the time of going to press. However, City & Guilds' products and services are subject to continuous development and improvement and the right is reserved to change products and services from time to time. City & Guilds cannot accept liability for loss or damage arising from the use of information in this publication.

City & Guilds
1 Giltspur Street
London EC1A 9DD

T 0844 543 0033
www.cityandguilds.com
publishingfeedback@cityandguilds.com

THE CITY & GUILDS TEXTBOOK

LEVEL 3 NVQ DIPLOMA IN ELECTROTECHNICAL TECHNOLOGY 2357

UNITS 305–306

JAMES L DEANS AND ANDREW HAY-ELLIS
SERIES EDITOR: PETER TANNER

ACKNOWLEDGEMENTS

City & Guilds would like to thank sincerely the following:

For invaluable knowledge and expertise
Eur Ing D. Locke, *BEng (Hons)*, *CEng*, *MIEE*, *MIEEE*, The Institution of Engineering and Technology, Technical Reviewer
Richard Woodcock, City & Guilds, Technical Reviewer and Contributor

For supplying pictures for the front cover and back cover
Jules Selmes and Adam Giles (photographer and assistant)

For their help with photoshoots
Jules Selmes and Adam Giles (photographer and assistant); Andrew Buckle (photographer); Andy Jeffery, Ben King and students from Oaklands College, St Albans; Andy Hay-Ellis, James L Deans, Jordan Hay-Ellis and the staff at Trade Skills 4 U, Crawley.

For help with pages 264–268
Roy Bater of Comptel Ltd.

Picture credits
Every effort has been made to acknowledge all copyright holders as below and the publishers will, if notified, correct any errors in future editions.

BSI p51, p69, p73, p87, p123, p138, p141, p142, p161 (permission to reproduce extracts from British Standards is granted by BSI Standards Limited (BSI). No other use of this material is permitted. British Standards can be obtained in PDF or hard copy formats from the BSI online shop: www.bsigroup.com/Shop or by contacting BSI Customer Services for hard copies only: Tel: +44 (0) 845 086 9001, Email: cservices@bsigroup.com); **Absorb Environmental Solutions** p156; **Axminster** p92, p94, p96, p97, p104, p116, p117; **BASEC** p58; **Batt Cables** p53; **Boaterbits** p112; **Castell Safety International Ltd** p29; **Ironmongery Direct** p111; **Kewtech** p34; **LockoutTagout.co.uk** p36, p145, p185; **Marshall-Tufflex** p66; **Martindale** p31, p33; **Pressmaster** p17; **Screwfix** p113, p117, p121; **Shutterstock** p1, p22, p26, p27, p42, p43, p49, p55, p57, p76, p78, p79, p90, p91, p92, p93, p94, p97, p99, p100, p102, p105, p106, p107, p112, p113, p114, p116, p118, p120, p140, p160, p165, p173, p174, p177, p192, p199, p200; **Unifix** p111

Text permissions
For kind permission of text extracts:

Crown copyright. Contains public sector information published by the Health and Safety Executive and licensed under the Open Government Licence v.1.0 and includes extracts from the following HSE publications: p10; p18; p19; p108, p181, p190, p191, p192, p193, p201

Permission to reproduce extracts from British Standards is granted by BSI Standards Limited (BSI). No other use of this material is permitted. British Standards can be obtained in PDF or hard copy formats from the BSI online shop: www.bsigroup.com/Shop or by contacting BSI Customer Services for hard copies only: Tel: +44 (0) 845 086 9001, Email: cservices@bsigroup.com; p22, p23, p24, p28, p29, p50, p61, p66, p82, p87, p89, p122, p123, p124, p130, p131, p133, p134, p136, p137, p141, p142, p144, p148, p149, p157, p159, p165, p178, p205

BS 7671: 2008 Incorporating Amendment No 1: 2011 can be purchased in hardcopy format only from the IET website http://electrical.theiet.org/ and the BSI online shop: http://shop.bsigroup.com

From the authors
James L Deans: Many thanks to my beautiful wife Jo-Ann and my two sons Hugo and Harley for all their patience and support.

Andrew Hay-Ellis: Many thanks to my wife, Michelle, for her patience and understanding during the many hours spent writing; also to my sons, Jordan and Brendan, for proofreading sections of text to make sure the words said what I intended.

CONTENTS

The regulations focus on making health and safety considerations a standard work activity when managing the project. It is intended that health and safety should be an integral part of the work and not be seen as just a bolt-on extra that only needs to be performed when someone is checking that work is being performed safely. In a modern organisation, the health, safety and environmental considerations control some of the most important issues and have a big impact on decision-making processes.

SAFE SITES

The key to achieving healthy and safe working conditions is to ensure that health and safety practices are planned and organised, to ensure that risks and hazards are controlled. The control measures must be monitored and reviewed. Everyone who is responsible for planning and controlling site work, as well as those working on the sites, has health and safety responsibilities. Checking that working conditions are healthy and the work environment is safe before work begins is essential. It is also vital to ensure that the proposed work is not going to put others at risk. This requires planning and organisation by the site management team.

It is important to remember that safety at work is a partnership between employer and employee and only by cooperating can we achieve this successfully. Site safety must be an attitude of mind and not just something that is done simply to 'tick the boxes'. A safe site:

- is safe to get in and out of, including in emergencies
- has ladders and scaffolding that are safe to use, with hand rails, etc
- has hazards fenced off, with clear warning signs
- is kept tidy
- has appropriate lighting
- has appropriate site security.

Persons involved in the planning and control of work activities on a site must be aware of the work requirements. This is one of the main reasons that, although risk assessments are the responsibility of the management team, they can be produced by anyone with a sound knowledge and understanding of the work activity.

It is important to gather as much health and safety information as possible about the project, the work and the proposed site before work begins. Information available at tendering should be used, to allow consideration of time and resources required to deal with particular problems. It is important that work methods and safety precautions are agreed and documented before work is started and that they are put into practice. It is essential to make sure everyone understands how work is to be done and is aware of relevant method statements before work starts.

ASSESSMENT GUIDANCE

Remember that low voltage goes up to 1000 V a.c.

Many people get confused because some lighting systems are referred to as low-voltage when in fact they are extra-low voltage (up to 50 V).

Assessment criteria

1.3 Specify the actions required to ensure that electrical installation work sites are correctly prepared in terms of health and safety considerations

ACTIVITY

How could site safety be maintained?

Permit to work schemes

One method of controlling work activity is to employ a Permit to Work, (P2W), scheme. A P2W enables a level of control over what is happening and when on any installation and can be employed throughout the life of the installation, i.e. during construction, use, maintenance and demolition.

For a P2W to work correctly the work activity must first have a risk assessment performed and **method statement** produced. These can in some cases be prepared by the person who will be responsible for performing the work activity, for example a main contractor will request the various sub-contracting trades to prepare their own risk assessments and method statements.

The actual permit to work is normally a paper document that is issued when a work activity is required. The permit will include copies of all associated risk assessments and method statements and will be issued to the person responsible for performing the task.

The person issuing the permit is responsible for ensuring that it is safe to follow the method statement and it is safe to perform the task. They must also be satisfied that the person to whom the permit is being issued is competent to receive the permit and perform the work activity. They must ensure that the permit is precise and accurate and must clearly identify when the permit will come into effect. It is important at this stage to ensure that the person who receives the permit understands the work required and the method to be used. Their confirmation is recorded on the form, which is signed by both the issuer and the recipient.

It is now the responsibility of the recipient to perform the task according the method statement and complete the task as required. They will then sign the permit to show that the task has been performed correctly, completely and absolutely. Once signed, the permit is returned to the issuer who will verify that the permit has been completed by the recipient.

On receipt of a completed permit the issuer will then cancel the permit so that it cannot be re-issued. This enables subsequent permits to be issued. By using the P2W system, a person can control all of the work activities on site, in accordance with the project plan, and be sure of the project progress and the safety of all persons on site. As people are not able to start work until they receive a P2W, this procedure ensures high-risk tasks can only be started when the necessary safety measures are in place. This is especially important when you have multiple trades working in the same area and their safety depends on everyone doing what they are supposed to, when they are supposed to.

Method statement

This is a written statement that identifies how the work activity will be performed and what safety measures will be employed to control the risks identified in the risk assessment, as well as detailing tools, equipment and personnel required. It may also detail other considerations such as access requirements, time frames and potential impacts of the work.

ASSESSMENT GUIDANCE

As the work progresses, risks may change due to your own or others' activities. Risks should be kept constantly under review.

Site preparation

It is often at the the planning stage of a work site that good foundations for health and safety are established. Several factors need to be considered.

Factor	Considerations include
Site access	■ How are vehicles to access the site to make deliveries and how are pedestrians going to be kept apart from vehicles to ensure safety? ■ How are access gates going to be controlled so that only authorised personnel are allowed onto the site? ■ How are visitors going to be kept safe during site visits? ■ How are members of the public, including children, going to be kept out?
Welfare facilities	■ How many toilets and wash rooms will be needed? ■ Where will toilets and wash rooms be placed? ■ Where will the canteen and rest facilities be positioned? ■ How many people will be expected to use the canteen and rest facilities at the same time? ■ How will wholesome water be made available for the workers on site?
Housekeeping	■ How will waste be disposed of? ■ What provision is needed for storage of tools and materials? ■ What access equipment will be required? ■ Where will access equipment be made secure when not in use?
Work environment	■ How will adequate on-site lighting be provided? ■ What provisions need to be made for power tools to be used on site? ■ What provisions need to be made for cleaning tools?
Emergency procedures	■ How will escape routes be identified and kept clear? ■ What emergency procedures will be required? ■ Where will fire assembly points be established? ■ How will fire registers be taken? ■ How will site personnel know there is an emergency incident? ■ How will site personnel know how to react to an emergency? ■ How many first-aiders will be required on site? ■ Where will the first-aid room be located? ■ What other emergency provisions will be required, including fire points for extinguishers? ■ Who else needs to be involved?
Site rules	■ How will the procedures and site rules be established? ■ Who needs to be involved in setting site rules? ■ How will site rules be communicated effectively to all site personnel? ■ How will site rules be monitored to ensure they are followed?

From the list, which is by no means exhaustive, it is clear that actually preparing a construction site takes a lot of effort and consideration, which is why it is important that everyone plays their part.

Pre-work checks

It is important that before you commence any work activity, you take the time to perform a few simple checks.

- Have you read the risk assessments for the work activity you are to perform?
- Have you read and understood the method statement for the work activity?
- Do you understand what the work entails and know how to perform the work safely and correctly?
- Are you able to perform the task safely?
- Do you have the correct tools, equipment and PPE?
- Do you have the relevant P2W or permissions to do the task?
- Is the permit active now?
- Is the location where the work activity is to be performed, clear of waste and other obstructions?
- Do you know where the nearest emergency exit is?
- Do you understand the site emergency procedures?

If you answer 'no' to any of these then you should not proceed with the task and should seek further guidance and advice from the site management team. Only when you are able to answer 'yes' to all of the above are you ready to perform the task.

OUTCOME 2

Understand the procedures for checking the work location prior to the commencement of work activities

PREPARATIONS PRIOR TO STARTING WORK

The previous chapter covered the need to prepare the work area and workforce to maintain health and safety during their work activities and time on site. This chapter covers the preparation work that needs to be performed with regards to the actual work activity.

With the site management team taking care of the working environment, it is normally left to either the sub-contractor, or their nominated team leader, to make the necessary preparations for the work activities that they will perform during the project. The preparation work is essential to performing the task correctly, cost effectively and, of course, safely. It is worth noting the five 'P's:

Proper

Planning

Prevents

Poor

Performance.

When an installation tender is submitted, it is based on getting it right first time and therefore making a profit. Every error, change or omission, no matter how small, has an impact on the final profit. Correct planning and preparation are intended to ensure that the work is performed quickly and correctly first time, every time. Another expression often used is; 'Fail to plan, plan to fail.' This is very true in electrical installation work.

Planning for materials and equipment

The most significant cost of any installation work is that of the labour required; however, a close second is that of the materials to be used. When confronted with a new project, it is essential to understand what materials will be required to complete the work. To help understand this, the installation team needs a number of documents.

The technical specification

This is the most important document as it has been used as the basis of the agreed contract that is being followed. However, it is important

Assessment criteria

2.1 State the preparations that should be completed before electrical installation work starts

SmartScreen Unit 305
Handout 6 and Worksheet 6

to ensure that all variations and departures are also documented and have been included. The technical specification will indicate any client's preferred manufacturers for certain components, or even desired finishes that need to be complied with.

The technical specification will also detail what other standards are to be complied with, as well as extra options required. For example, the client may require a minimum of 25% of extra capacity on every distribution board. This would have an impact on the equipment ordered. For example, if a distribution board was specified as providing a supply to ten circuits, then typically a 16-**way** distribution board would be required to meet the client's request.

Other key information that is normally contained within the technical specification is the detail of the building construction. The technical specification will detail what the wall construction is, along with the details of the floor, wall and ceilings finishes. This information is essential when planning an electrical installation, as it helps to identify where and how the wiring can be installed. If an installation team were not aware that the floor through which they were expecting to run cables was actually a concrete slab, then this would create a lot of problems on site and would result in a significant delay if it was not discovered quickly.

It is also important to consider other factors such as the purpose of the building, who will be using it and other environmental factors. These may be detailed within the technical specification but, if not, then the design team need to establish these external influences. BS7671 dedicates an appendix to these very influences.

Site drawings

The site drawings will include this information.

Drawing	Details
Site plans	These will provide the details of where large components are to be sited and will often have either multiple sheets for the different services or a series of overlays.
Layout diagrams	These will show the positioning of all components to be installed. If components are to be mounted at heights other than those covered in part M of the building regulations, then this information is provided as well.
Cable routes	These plans will show the proposed routes of cables throughout the installation and these can be used for establishing approximate cable lengths to enable cables to sized and specified.

Way

When used in relationship to distribution boards, the way is the maximum number of single-phase circuits that can be connected to the board.

ASSESSMENT GUIDANCE

Do not assume that every floor of a building is constructed in the same way, or that each house on an estate is the same. Even something such as finding hardboard under a carpet can cause unexpected delays and expense.

KEY POINT

Part M of the building regulations specifies that electrical accessories that a user in a wheelchair may be required or expected to operate must be located in an area where such a person can easily reach. This relates to the zone 450 mm from finished floor level (ffl), to 1200 mm ffl. If the installation is to be used by a person in a wheelchair, then this must be taken into account in relation to heights and ease of operation.

Using site drawings to plan the job

By using all of the available site drawings it is possible to get an overview of all the materials that will be required to complete the installation work.

Materials list

Trying to keep track of all the materials required, just by marking up drawings, can become very complicated, especially on larger projects. Therefore it is common practice to use a series of documents generically called the materials list.

Materials lists will contain different information, depending on company procedures and requirements, however there are always some common aspects and these include:

- the index number – a simple numerical identifier that makes it easier to track items on the materials list
- the description –a brief description of the item to enable it to be identified
- the unit of measure –used to clarify what the quantities are measured in, for example, litres, metres or individual items
- the quantity – how many of the items are required.

ASSESSMENT GUIDANCE

The method of material procurement will depend upon your working arrangements and even finance available. Are you going to rely on the wholesaler always having the materials in stock, or will you keep some items yourself to meet out of hours requirements.

Electrical buyer

The person who places the purchase orders with suppliers for electrical equipment. They are often required to source and negotiate deals to secure the best price and delivery for a wide selection of electrical accessories and components. Big discounts can be given by the supplier if the electrical buyer can place large orders.

ACTIVITY

A contractor requires 24 lengths of 20 mm galvanised conduit. The price per length is £6.99 bought individually, but when all 24 are purchased together, the total price is £142.80. What is the percentage discount?

Short shipments

These occur when the supplier delivers an incomplete order. There are normally items on the original order still to be delivered, or the quantity is short and the balance is yet to come.

Just in time

A large-quantity order is placed with a supplier, with agreement on a staged delivery. This means that the total order is placed at the start, to take advantage of quantity discounts, but the supplier stores the materials and only delivers the required quantity at the agreed times throughout the project.

Other information that may appear on a materials list includes:

- cost – how much each item costs
- supplier – who is the current supplier or who stocks the item
- catalogue code – where a specific item must be used, this is the suppliers catalogue number.

Materials lists can be arranged in two different formats, depending on who is to use the information. **Electrical buyers** may want to know the total counts of materials so that they can exploit a higher level of buying power. For example, if an electrical buyer needs to order several thousand components from the same supplier, then they will be able to request a bigger discount.

An installation team will require the materials list to be broken down, to show what they would need in each part of the installation. This enables them undertake stock control and make sure that the correct quantities of materials are in place at the right time. Being able to do this means that there is less waste of materials and other losses such as delays and **short shipments** are kept to a minimum.

Working from materials lists can enable bulk purchasing power to be used to gain a lower cost. It can also be used to set delivery schedules with the suppliers using **just in time** systems. This means that there is less requirement for storage of materials on site and less chance of materials being lost, damaged or even stolen.

Tools and equipment

With the design completed, risk assessments performed, with method statements produced and all materials identified and ordered, the final stage of the pre-construction preparation is about ensuring that the correct tools are available, safe and in good working condition. In reviewing the site drawings it will also have become apparent what access equipment and working platforms will be required to perform the installation work safely.

The Provision and Use of Work Equipment Regulations, (PUWER), state under Regulation 4 that:

'(1) Every employer shall ensure that work equipment is so constructed or adapted as to be suitable for the purpose for which it is used or provided.

(2) In selecting work equipment, every employer shall have regard to the working conditions and to the risks to the health and safety of persons which exist in the premises or undertaking in which that work equipment is to be used and any additional risk posed by the use of that work equipment.'

All tools must be checked at this point, to make sure that they are safe to be used. Any defective tools or equipment must be replaced

ELECTRICAL INSTALLATION CONDITION REPORT

SECTION A. DETAILS OF THE CLIENT / PERSON ORDERING THE REPORT

Name ....

Address ....

SECTION B. REASON FOR PRODUCING THIS REPORT ....

Date(s) on which inspection and testing was carried out ....

SECTION C. DETAILS OF THE INSTALLATION WHICH IS THE SUBJECT OF THIS REPORT

Occupier ....

Address ....

Description of premises (tick as appropriate)

Domestic ☐   Commercial ☐   Industrial ☐   Other (include brief description) ☐

Estimated age of wiring system ....years

Evidence of additions / alterations   Yes ☐   No ☐   Not apparent ☐   If yes, estimate age ....years

Installation records available? (Regulation 621.1)   Yes ☐   No ☐   Date of last inspection .... (date)

SECTION D. EXTENT AND LIMITATIONS OF INSPECTION AND TESTING

Extent of the electrical installation covered by this report ....

Agreed limitations including the reasons (see Regulation 634.2) ....

Agreed with: ....

Operational limitations including the reasons (see page no. ....) ....

The inspection and testing detailed in this report and accompanying schedules have been carried out in accordance with BS 7671: 2008 (IET Wiring Regulations) as amended to ....

It should be noted that cables concealed within trunking and conduits, under floors, in roof spaces, and generally within the fabric of the building or underground, have **not** been inspected unless specifically agreed between the client and inspector prior to the inspection.

SECTION E. SUMMARY OF THE CONDITION OF THE INSTALLATION

General condition of the installation (in terms of electrical safety) ....

Overall assessment of the installation in terms of its suitability for continued use

SATISFACTORY / UNSATISFACTORY* (Delete as appropriate)

*An unsatisfactory assessment indicates that dangerous (code C1) and/or potentially dangerous (code C2) conditions have been identified.

SECTION F. RECOMMENDATIONS

Where the overall assessment of the suitability of the installation for continued use above is stated as UNSATISFACTORY, I / we recommend that any observations classified as *Danger present* (code C1) or *Potentially dangerous* (code C2) are acted upon as a matter of urgency.

Investigation without delay is recommended for observations identified as *further investigation required*.

Observations classified as *Improvement recommended* (code C3) should be given due consideration.

Subject to the necessary remedial action being taken, I / we recommend that the installation is further inspected and tested by .... (date)

SECTION G. DECLARATION

I/We, being the person(s) responsible for the inspection and testing of the electrical installation (as indicated by my/our signatures below), particulars of which are described above, having exercised reasonable skill and care when carrying out the inspection and testing, hereby declare that the information in this report, including the observations and the attached schedules, provides an accurate assessment of the condition of the electrical installation taking into account the stated extent and limitations in section D of this report.

Inspected and tested by:	Report authorised for issue by:
Name (Capitals)	Name (Capitals) ....
Signature	Signature ....
For/on behalf of	For/on behalf of ....
Position	Position ....
Address	Address ....
Date	Date ....

SECTION H. SCHEDULE(S)

..........schedule(s) of inspection and ....schedule(s) of test results are attached.

The attached schedule(s) are part of this document, and this report is valid only when they are attached to it.

SECTION I. SUPPLY CHARACTERISTICS AND EARTHING ARRANGEMENTS

Earthing arrangements	Number and Type of Live Conductors	Nature of Supply Parameters	Supply Protective Device
TN-C ☐	a.c. ☐ d.c. ☐	Nominal voltage, $U / U_0^{(1)}$ V	BS (EN) ....
TN-S ☐	1-phase, 2-wire ☐ 2-wire ☐	Nominal frequency, $f^{(1)}$ .... Hz	Type ....
TN-C-S ☐	2 phase, 3-wire ☐ 3-wire ☐	Prospective fault current, $I_{pf}^{(2)}$ .... kA	Rated current ....A
TT ☐	3 phase, 3-wire ☐	External loop impedance, $Ze^{(2)}$ .... Ω	
IT ☐	3 phase, 4-wire ☐	Note: (1) by enquiry	
	Confirmation of supply polarity ☐	(2) by enquiry or by measurement	

Other sources of supply (as detailed on attached schedule) ☐

SECTION J. PARTICULARS OF INSTALLATION REFERRED TO IN THE REPORT

Means of Earthing **Details of Installation Earth Electrode** (*where applicable*)

Distributor's facility ☐   Type ....

Installation earth electrode ☐   Location ....

Resistance to Earth ....Ω

Main Protective Conductors

Earthing conductor ☐	Material	Csa ....mm²	Connection / continuity verified ☐
Main protective bonding conductors ☐	Material	Csa ....mm²	Connection / continuity verified ☐

To incoming water service ☐   To incoming gas service ☐   To incoming oil service ☐   To structural steel ☐

To lightning protection ☐   To other incoming service(s) ☐   Specify ☐

Main Switch / Switch-Fuse / Circuit-Breaker / RCD

Location	Current rating ....A	**If RCD main switch**
BS(EN)	Fuse / device rating or setting ....A	Rated residual operating current ($I_{\Delta n}$) ....mA
No of poles	Voltage rating V	Rated time delay ....ms
		Measured operating time(at $I_{\Delta n}$) ....ms

SECTION K. OBSERVATIONS

Referring to the attached schedules of inspection and test results, and subject to the limitations specified at the *Extent and limitations of inspection and testing* section

No remedial action is required ☐   The following observations are made ☐ (see below):

OBSERVATION(S)	CLASSIFICATION CODE	FURTHER INVESTIGATION REQUIRED (YES / NO)

One of the following codes, as appropriate, has been allocated to each of the observations made above to indicate to the person (s) responsible for the installation the degree of urgency for remedial action.

C1 – Danger present. Risk of injury. Immediate remedial action required

C2 – Potentially dangerous - urgent remedial action required

C3 – Improvement recommended

Electrical Installation Condition Report (always check you are using the latest forms, as found on the IET website: http://electrical.theiet.org

CONDITION REPORT INSPECTION SCHEDULE FOR DOMESTIC AND SIMILAR PREMISES WITH UP TO 100 A SUPPLY

Note: This form is suitable for many types of smaller installation not exclusively domestic.

OUTCOMES: Acceptable condition ✓ | Unacceptable condition : State C1 or C2 | Not verified: NV | Limitation :LIM | Not applicable N/A

Outcome columns: State C1 or C2 | Improvement recommended : State C3 | Not verified: NV | Limitation: LIM | Not applicable | OUTCOME (Use codes above. Provide additional comment where appropriate. C1, C2 and C3 coded items to be recorded in Section K of the Condition Report) | Further investigation required? (Y or N)

ITEM NO	DESCRIPTION
1.0	**DISTRIBUTOR'S / SUPPLY INTAKE EQUIPMENT**
1.1	Service cable condition
1.2	Condition of service head
1.3	Condition of tails - Distributor
1.4	Condition of tails - Consumer
1.5	Condition of metering equipment
1.6	Condition of isolator (where present)
2.0	**PRESENCE OF ADEQUATE ARRANGEMENTS FOR OTHER SOURCES SUCH AS MICROGENERATORS (551.6; 551.7)**
3.0	**EARTHING / BONDING ARRANGEMENTS (411.3; Chap 54)**
3.1	Presence and condition of distributor's earthing arrangement (542.1.2.1; 542.1.2.2)
3.2	Presence and condition of earth electrode connection where applicable (542.1.2.3)
3.3	Provision of earthing / bonding labels at all appropriate locations (514.11)
3.4	Confirmation of earthing conductor size (542.3; 543.1.1)
3.5	Accessibility and condition of earthing conductor at MET (543.3.2)
3.6	Confirmation of main protective bonding conductor sizes (544.1)
3.7	Condition and accessibility of main protective bonding conductor connections (543.3.2; 544.1.2)
3.8	Accessibility and condition of all protective bonding connections (543.3.2)
4.0	**CONSUMER UNIT(S) / DISTRIBUTION BOARD(S)**
4.1	Adequacy of working space / accessibility to consumer unit / distribution board (132.12; 513.1)
4.2	Security of fixing (134.1.1)
4.3	Condition of enclosure(s) in terms of IP rating etc (416.2)
4.4	Condition of enclosure(s) in terms of fire rating etc (526.5)
4.5	Enclosure not damaged/deteriorated so as to impair safety (621.2(iii))
4.6	Presence of main linked switch (as required by 537.1.4)
4.7	Operation of main switch (functional check) (612.13.2)
4.8	Manual operation of circuit-breakers and RCDs to prove disconnection (612.13.2)
4.9	Correct identification of circuit details and protective devices (514.8.1; 514.9.1)
4.10	Presence of RCD quarterly test notice at or near consumer unit / distribution board (514.12.2)
4.11	Presence of non-standard (mixed) cable colour warning notice at or near consumer unit / distribution board (514.14)
4.12	Presence of alternative supply warning notice at or near consumer unit / distribution board (514.15)
4.13	Presence of other required labelling (please specify) (Section 514)
4.14	Examination of protective device(s) and base(s); correct type and rating (no signs of unacceptable thermal damage, arcing or overheating) (421.1.3)
4.15	Single-pole protective devices in line conductor only (132.14.1; 530.3.2)
4.16	Protection against mechanical damage where cables enter consumer unit / distribution board (522.8.1; 522.8.11)
4.17	Protection against electromagnetic effects where cables enter consumer unit / distribution board enclosures (521.5.1)
4.18	RCD(s) provided for fault protection – includes RCBOs (411.4.9; 411.5.2; 531.2)
4.19	RCD(s) provided for additional protection - includes RCBOs (411.3.3; 415.1)

ITEM NO	DESCRIPTION
5.0	**FINAL CIRCUITS**
5.1	Identification of conductors (514.3.1)
5.2	Cables correctly supported throughout their run (522.8.5)
5.3	Condition of insulation of live parts (416.1)
5.4	Non-sheathed cables protected by enclosure in conduit, ducting or trunking (521.10.1)
	• To include the integrity of conduit and trunking systems (metallic and plastic)
5.5	Adequacy of cables for current-carrying capacity with regard for the type and nature of installation (Section 523)
5.6	Coordination between conductors and overload protective devices (433.1; 533.2.1)
5.7	Adequacy of protective devices: type and rated current for fault protection (411.3)
5.8	Presence and adequacy of circuit protective conductors (411.3.1.1; 543.1)
5.9	Wiring system(s) appropriate for the type and nature of the installation and external influences (Section 522)
5.10	Concealed cables installed in prescribed zones (see Section D. Extent and limitations) (522.6.101)
5.11	Concealed cables incorporating earthed armour or sheath, or run within earthed wiring system, or otherwise protected against mechanical damage from nails, screws and the like (see Section D. Extent and limitations) (522.6.101; 522.6.103)
5.12	Provision of additional protection by RCD not exceeding 30 mA:
	• for all socket-outlets of rating 20 A or less provided for use by ordinary persons unless an exception is permitted (411.3.3)
	• for supply to mobile equipment not exceeding 32 A rating for use outdoors (411.3.3)
	• for cables concealed in walls or partitions (522.6.102; 522.6.103)
5.13	Provision of fire barriers, sealing arrangements and protection against thermal effects (Section 527)
5.14	Band II cables segregated / separated from Band I cables (528.1)
5.15	Cables segregated / separated from communications cabling (528.2)
5.16	Cables segregated / separated from non-electrical services (528.3)
5.17	Termination of cables at enclosures – indicate extent of sampling in Section D of the report (Section 526)
	• Connections soundly made and under no undue strain (526.6)
	• No basic insulation of a conductor visible outside enclosure (526.98)
	• Connections of live conductors adequately enclosed (526.5)
	• Adequately connected at point of entry to enclosure (glands, bushes etc.) (522.8.5)
5.18	Condition of accessories including socket-outlets, switches and joint boxes (621.2 (iii))
5.19	Suitability of accessories for external influences (512.2)
6.0	**LOCATION(S) CONTAINING A BATH OR SHOWER**
6.1	Additional protection for all low voltage (LV) circuits by RCD not exceeding 30 mA (701.411.3.3)
6.2	Where used as a protective measure, requirements for SELV or PELV met (701.414.4.5)
6.3	Shaver sockets comply with BS EN 61558-2-5 formally BS 3535 (701.512.3)
6.4	Presence of supplementary bonding conductors, unless not required by BS 7671:2008 (701.415.2)
6.5	Low voltage (e.g. 230 volt) socket-outlets sited at least 3 m from zone 1 (701.512.3)
6.6	Suitability of equipment for external influences for installed location in terms of IP rating (701.512.2)
6.7	Suitability of equipment for installation in a particular zone (701.512.3)
6.8	Suitability of current-using equipment for particular position within the location (701.55)
7.0	**OTHER PART 7 SPECIAL INSTALLATIONS OR LOCATIONS**
7.1	List all other special installations or locations present, if any. (Record separately the results of particular inspections applied.)

Inspected by:
Name (Capitals) .... Signature .... Date ....

Condition Report Inspection Schedule(s) (always check you are using the latest forms, as found on the IET website: http://electrical.theiet.org)

The purpose of the camera is to make sure that there is documented proof of the condition prior to work commencing, should any problems or damage be identified. Simply taking a picture of the damage or problem can often help to clarify with the client what work will be undertaken. It is important to focus on the area where the work is going to be performed, including access routes.

If a pre-work survey is not conducted and the client notices some damage to their property after the work is finished, they may blame the installer for the damage and expect it to be corrected, with the installer footing the bill, which can be expensive. If, however, a pre-work survey has been performed, this may prove that the installer is not responsible for the damage.

KEY POINT

Using your phone as a camera is acceptable, but remember to print the photos so that that they can be preserved.

What to look for

As well as checking the condition of the electrical installation, it is important to check the surrounding areas and access routes for existing damage, rather like checking a hire car before driving it away. Checking for existing damage is important as it:

- may identify challenges for the installer that had not been noticed before
- can reveal potential problems within the installation
- can identify where further work may be required.

These are some of the things that should be checked for outside of the installation.

Damage to:	Look for
the actual building in terms of its structure	- cracks along walls and ceilings - dents in walls and ceilings - holes in walls and ceilings - loose fittings caused by crumbling plaster.
decorations such as wall paper	- torn or peeling paper - marks on wall paper or painted walls - chips to paintwork or other marks.
furniture and fittings	- cuts, holes or burns on furniture - water marks on wooden surfaces - heat rings on wooden surfaces.

Damage to:	Look for
carpets and flooring	■ dirty marks ■ stains from spills or paint ■ holes, burns or cuts ■ scratches on wooden floors.
doors, windows and their frames	■ broken or chipped glass ■ broken handles or locks ■ chips or cracks to woodwork ■ existing holes or notches.

This list is intended to serve only as a starting point for consideration. It is important that, before work begins and before tools and accessories are brought into the installation, time is taken to consider the surroundings. This does not need to take a lot of time, but be mindful of checking thoroughly all areas to be accessed and worked in.

What to do if damage is discovered

If damaged is discovered during the pre-work survey then the installer simply needs to make a note of the damage and take a quick picture. Then, before starting work, the installer should speak to the client and point out the damage identified. A lot of tact is required, as the installer will potentially be pointing out problems with the client's property.

It is always important to make the client aware that the damage is only being identified to confirm they are already aware of its existence.

If damage is discovered during work then it is important again for the installer to record it and, where possible, take a photograph of the damage. If the damage is severe then the client also needs to be made aware of the damage as soon as possible. This may mean that work has to stop, depending on the severity of the damage. This typically includes damage to structural elements, such as floor joists that were not visible before the work commenced.

Damaged cables

Another type of damage that might be identified at this time, rather than earlier, is damage to wiring caused by overheating conductors or rodents chewing cables. In this instance it is important to show this to the client as soon as is practicable, as it may require further action that is beyond the scope of the planned work. Damaged wiring presents an immediate risk of electric shock. It must be made safe immediately and this may involve safely isolating the circuit and getting guidance from the client as to the remedial action they require.

This often will require the client to speak with the company employing the installer, to agree a contract detailing additional work is to be performed and its cost.

PROTECTING THE PROPERTY

After checking for existing damage prior to starting work, it is just as important to make sure that damage does not occur as the work is being carried out. Precautions must be taken to prevent damage from occurring whenever possible. To prevent damage to key items, some simple measures can be taken, such as:

- wearing overshoes to protect the client's carpets
- using floor matting to protect the client's flooring
- laying dust sheets to protect the client's furnishings
- erecting hoardings to protect windows and other areas of the property.

Wearing overshoes

Where the work areas house small items of valuable property belonging to the client, it is a good idea to ask the client if they are happy to remove them so that they do not get damaged. Doing this removes the risk of the items being damaged, either during the work process or even as they are being moved by the installer.

KEY POINT

An immediate risk of electric shock is present if live parts can be touched due to a failing of basic protection.

Assessment criteria

2.4 Specify methods for protecting the fabric and structure of the property before and during installation work

ACTIVITY

Carrying out a domestic rewire will be easier if the materials and tools can be stored in a spare room or garage. Name some suitable alternatives.

OUTCOME 3

Understand the practices, procedures and regulatory requirements for completing the safe isolation of electrical circuits and complete electrical installations

ISOLATION PROCEDURE

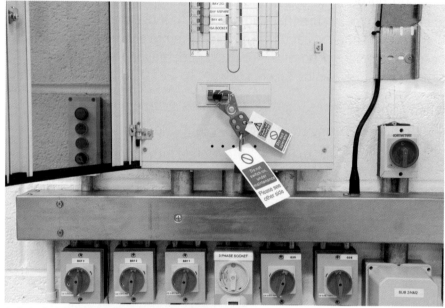

Safe isolation includes using notices like these

A procedure is a series of steps or actions, taken to accomplish an outcome. Safe isolation procedure requires that a series of actions is carried out in a set order.

Reasons for isolation

Isolation is defined in BS 7671 as:

> 'a function intended to cut off for reasons of safety the supply from all, or a discrete section, of the installation by separating the installation or section from every source of electrical energy'.

The primary reason for isolation is to remove the risk of electric shock, fire and burns when working on the electrical system.

The Electricity at Work regulations 1989 take this a step further, as Regulation 13 states:

> 'Adequate precautions shall be taken to prevent electrical equipment, which has been made dead in order to prevent danger

while work is carried out on or near that equipment, from becoming electrically charged during that work if danger may thereby arise.'

BS 7671 provides further requirements, under Section 537, that also need to be understood when considering devices for isolation. Regulation 537.2.1.5:

'Where an isolating device for a particular circuit is placed remotely from the equipment to be isolated, provision shall be made so that the means of isolation can be secured in the open position. Where this provision takes the form of a lock or removable handle, the key or handle shall be non-interchangeable with any other used for a similar purpose within the premises.'

Regulation 537.2.2.3:

'A device for isolation shall be designed and/or installed so as to prevent unintentional or inadvertent **closure**.'

When using a switch to perform an isolating function, the most common way of achieving this is to ensure the switch can be locked in the 'OFF' position. Another option is to use a key exchange system that only enables certain switches to be operated by specially designed keys that are unique and can only be released when other unique keys are inserted in an exchange lock system.

The impact of failure to carry out safe isolation

People often think, incorrectly, that failure to isolate an installation safely only affects the person working on the electrical installation. The table shows, however, that the effects of failure to isolate safely can be far reaching.

An example of a key exchange system and lock

> **KEY POINT**
>
> The use of the word 'safe' when referring to isolation means that the requirements of regulation 13 from the Electricity at Work regulations 1989 must be complied with.

Closure

A switch is typically referred to as closed when it is switched on and open when switched off.

> **ASSESSMENT GUIDANCE**
>
> Carry out the safe isolation yourself and ensure that all the necessary power is disconnected. Do not rely on someone else.

Areas or people affected	Hazards to consider
Operative working on the system	■ Risk of electric shock when working on system
Other personnel	■ Risk of electric shock from contact with live parts when barriers are removed
Customers and clients	■ Risk of electric shock ■ Damage to equipment if short circuits are created when working on live equipment
General public	■ Risk of electric shock
Building services	■ Short circuits created when working on live equipment may damage building systems equipment, resulting in damage to the building fabric due, for example, to heat.

The impact of carrying out isolation

Although safe isolation removes the risk of electric shock, the act of isolation can have a negative impact on other users of the installation and, in some cases, can give rise to additional hazards. Careful planning of isolation can minimise the risk of harm from these hazards. The table indicates some of the possible hazards that should be considered when planning an isolation.

Areas or people affected	Hazards to consider
Other personnel	■ Unexpected loss of power to machines, giving rise to dangerous situations ■ Loss of lighting which may be required to carry out normal operations safely
Customers and clients	■ Loss of service will affect normal operations, for instance, in a retail environment, the loss of supply to tills and card machines will stop sales being made
General public	■ Loss of lighting, emergency lighting and fire alarms (both fire alarms and emergency lighting systems have battery back-up systems but these deplete over time, so the length of time the power is isolated must be minimised)
Building services	■ Unexpected loss of power to computers, resulting in data loss ■ Loss of communications systems ■ Loss of services such as heating and ventilation ■ Some processes may require a shut-down procedure and a restart procedure when the supply is reinstated ■ Loss of supply to lifts meaning people may be trapped

ACTIVITY

When working in a house, shop or office where only a single circuit can be isolated and others must remain live, how could this single circuit be safely isolated?

Before carrying out isolation, it is important that any hazards that may arise due to the isolation are identified. A risk assessment should be carried out and a method statement should be agreed. It is imperative that permission is gained from the person responsible for the building's operation *prior* to carrying out isolation. It is important to consider other potential hazards that may not be directly related to the work involved. These include:

3 Confirm that the device used for isolation is suitable and may be secured effectively

and

4 Power down circuit loads if isolator is not suitable for on-load switching

When deciding on the suitability of a means of isolation, ask these questions.

- Can the means of isolation interrupt the circuit under full load conditions?
- Can the means of isolation be locked in the off position?
- Is it clear to see that isolation has taken place? Are the 'on' and 'off' positions clearly marked?

Not all means of isolation are designed to interrupt the circuit under load conditions. It is good practice to switch off or unplug individual loads before operating the means of isolation. When re-energising, the reverse process is followed: switch on the means of isolation, then reconnect individual loads one by one.

If the means of isolation cannot be secured with a padlock, other methods of securing the isolation must be used, such as withdrawal of fuses or disconnection of circuit conductors, or even removal of plug from socket.

Some functional switches, such as light switches, cannot be used as a means of isolation as their 'off' positions are not clearly and reliably indicated and they cannot be secured against unintentional re-energisation.

> **KEY POINT**
>
> If a plug and socket are being used as a means of isolation, then the plug must be under the constant supervision of the person performing the work. If it cannot be, then alternative means of isolation must be employed.

5 Disconnect, using the located isolator (from step 2)

Disconnection may be achieved by:

- switching the isolation device or switch disconnector to the off position
- withdrawal of circuit fuses
- operation of circuit breakers or RCDs to the off position
- unplugging of equipment
- disconnection of circuit conductors.

Operation of functional switches, such as light switches, is not a suitable method of isolation.

Whichever method is used, it is important that the equipment or circuit cannot be accidentally re-energised.

6 Secure in the off position, keep key on person, and post warning signs

A fundamental principle of isolation is that the means of isolation must be under the control of the person carrying out the work on the isolated circuit or installation, so that the installation cannot be accidentally re-energised. The ideal way of ensuring this is to use a lock with a unique key and for the operative to keep the key in their pocket.

With switchgear, it may be a case of fitting a padlock; with circuit breakers a proprietary lock-off device may be required. Manufacturers of circuit breakers make these devices to suit their particular products. Universal lock-off devices are available and it is a good idea to have a varied supply to hand.

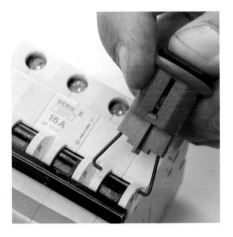
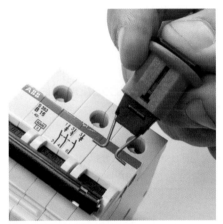

Various lock-off devices

Secured supply

On larger installations, more than one person may be working at the same time. Multiple lock-off devices are available specifically for these situations. Using one of these devices ensures the installation is not re-energised until everyone has finished and removed their lock.

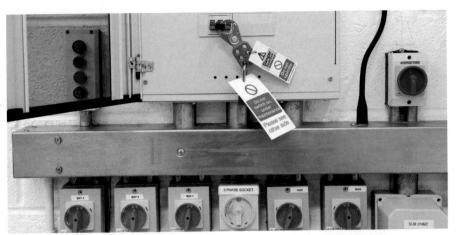

Multiple lock-off device

In the case of a fuse board without the facility to install a lock, the fuses should be withdrawn and secured so that they cannot be reinserted until work has finished. 'Securing the fuses' may simply mean that they are put in the operative's pocket or toolbox; better still, they can be locked in a purpose-designed box.

In some cases the only option may be to disconnect the circuit conductors of the circuit supplying the equipment being worked on. In either case it is important to label the fuse board or circuit conductors so others know why the fuses have been removed or the circuit has been disconnected.

In some cases the means of disconnection may be simply the withdrawal of a plug. Lock-off devices do exist to enable operatives to ensure that the plug is not unintentionally reinserted.

Search the internet to locate suitable lock-off devices for the situations described above. Whatever method is used, isolation must remain secure at all times.

7 Using voltage indicator, confirm isolation by checking ALL combinations

and

8 Prove voltage indicator on known source, such as proving unit

You must prove the voltage indicator is working before and after using it to confirm that the equipment or circuit is really 'dead'. You can use a known source, for instance, if the switch disconnector on a distribution board has been used as the means of isolation, one side will be live and the other side should be dead. The live side can be used as the known source.

If a known source is not available, use a proving unit. A proving unit is a battery source, with a high-voltage output at low current, that is suitable for checking the correct operation of a voltage indicator. Proving units *do not* work with test lamps as these require a higher current to operate, so the only option here is to test on a known source.

Proving unit

It is very important that all combinations of conductors are checked.

- For single phase, three individual tests are used (known as the three-point test):

1	Line	Neutral	
2	Line		Earth
3		Neutral	Earth

- For three-phase equipment or circuits, there are ten individual tests (known as the 10-point test):

1	Line 1	Line 2			
2	Line 1		Line 3		
3	Line 1			Neutral	
4	Line 1				Earth
5		Line 2	Line 3		
6		Line 2		Neutral	
7		Line 2			Earth
8			Line 3	Neutral	
9			Line 3		Earth
10				Neutral	Earth

Never omit any of the tests, even the neutral–earth test is important. It is not uncommon to find installations where the neutral is from one circuit and the line from another. This incorrect practice is known as 'borrowed' or 'shared neutrals' and is highly dangerous; isolating the line conductor could leave the neutral at a dangerous potential. The neutral–earth test identifies such circuits.

Place warning notices

Warning notices advise other people about what has been done. A notice on an isolator informs others that the operative has switched off because they are working on the installation and that the isolation has taken place for safety reasons.

Warning notice

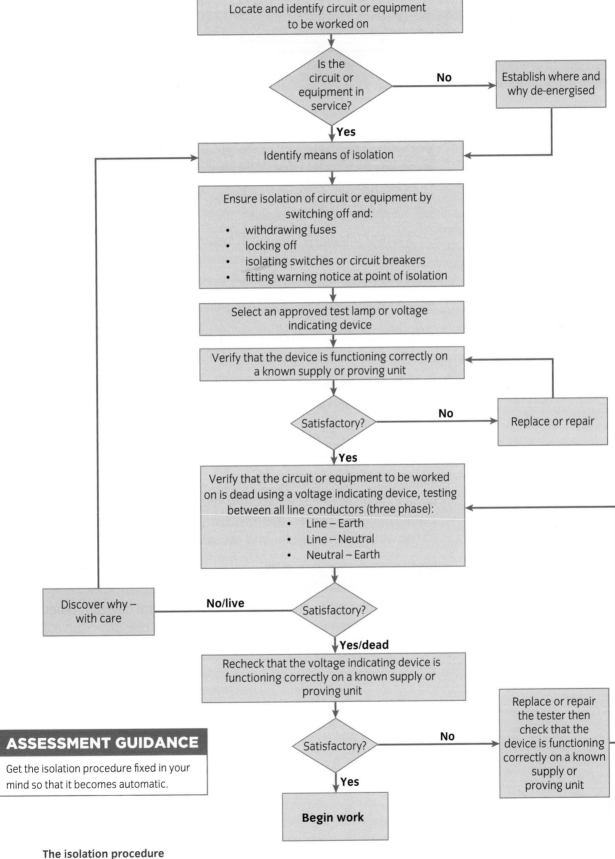

Isolation procedure

Locate and identify circuit or equipment to be worked on

Is the circuit or equipment in service? — **No** → Establish where and why de-energised

Yes

Identify means of isolation

Ensure isolation of circuit or equipment by switching off and:
- withdrawing fuses
- locking off
- isolating switches or circuit breakers
- fitting warning notice at point of isolation

Select an approved test lamp or voltage indicating device

Verify that the device is functioning correctly on a known supply or proving unit

Satisfactory? — **No** → Replace or repair

Yes

Verify that the circuit or equipment to be worked on is dead using a voltage indicating device, testing between all line conductors (three phase):
- Line – Earth
- Line – Neutral
- Neutral – Earth

Discover why – with care ← **No/live** — Satisfactory?

Yes/dead

Recheck that the voltage indicating device is functioning correctly on a known supply or proving unit

Satisfactory? — **No** → Replace or repair the tester then check that the device is functioning correctly on a known supply or proving unit

Yes

Begin work

The isolation procedure

ASSESSMENT GUIDANCE

Get the isolation procedure fixed in your mind so that it becomes automatic.

OUTCOME 4

Understand the types, applications and limitations of wiring systems and associated equipment

Assessment criteria

4.1 State the constructional features, applications, advantages and limitations of types of cable

4.2 State the constructional features, applications, advantages and limitations of types of cable and conductor containment systems

4.3 Describe how environmental factors can affect the selection of wiring systems, associated equipment and enclosures

4.4 State the types of wiring systems and associated equipment used for different types of system

SELECTION OF WIRING SYSTEMS

The Electricity at Work Regulations 1989, Regulation 2 defines an electrical system as:

> '"system" means an electrical system in which all the electrical equipment is, or may be, electrically connected to a common source of electrical energy, and includes such source and such equipment.'

BS7671 has a similar definition in that it defines a system as:

> 'An electrical system comprising of a single source or multiple sources running in parallel of electrical energy and an installation.'

Both definitions identify the installation as part of the system and an installation is defined in BS7671 as:

> 'An assembly of associated electrical equipment having coordinated characteristics to fulfil specific purposes.'

When reference is made to a system it can include the wiring, containment, accessories and the source of electrical energy. For the purpose of this chapter, the term 'wiring system' shall be used to refer to the wiring, containment and accessories only.

Assessment criteria

4.1 State the constructional features, applications, advantages and limitations of types of cable

TYPES OF CABLE

In the electrotechnical industry there are several different types of cable that are used for different purposes and reasons. 'Cable' is normally the term used to describe a grouping of wires into one item.

The parts of a cable

Sheath	This is the outer covering and is not only used to hold the wires together in a group but also provides mechanical protection to the insulation of the wires within.
Insulation	This is the covering around the conductor and performs two functions. The first is that of basic protection to stop persons from coming into contact with live parts. The second function is one of identification and to this extent it is common for insulation to be colour coded. BS7671 Appendix 7 gives details of the harmonised colour coding that can be found in cables with regards to their insulations.
Conductor	This is the part that carries the electrical energy. This is typically made out of copper but can also be made from aluminium. Where the conductor is made from copper, BS 7671 regulation 524.1 identifies a minimum cross-sectional area of 1.0 mm^2 for lighting and 1.5 mm^2 for power circuits. However, aluminium has a minimum sizing of 16 mm^2 for all types of circuit.

The insulation material used is normally rated in terms of dielectric strength. This relates to how well the insulation can withstand an electric field without being damaged or breaking down. This is based on the fact that all materials will conduct, but at different levels of electric field strength. The simple way to consider this is that air can be used as an insulator but lightning passes through air with considerable ease.

Wiring is also sometimes referred to by how the conductors are formed. Conductors can be formed in three main ways.

KEY POINT

An insulator with a high dielectric loss should not be used on high-voltage cables.

Solid	The conductor is made of a single piece of conductive material and this is generally round until the conductor cross-sectional area (CSA) reaches larger sizes, typically 300 mm^2 and higher, after which they are shaped to save on wasted space in the cable assembly.
Stranded	Conductors made of solid conductive materials tend to be hard to bend once they get above 2.5 mm^2 and therefore, to make it easier to bend, the conductor is normally made up of a collection of smaller strands of conductive material, all held together by the insulation. Consequently, the overall size of the cable is increased. Once the cable reaches a CSA where it is unlikely to have many bends then the conductors are made solid again.
Flex	When the wire of a cable is expected to be subjected to vibration or regular movement while in use, BS 7671 states under regulations 521.9.2 and 521.9.3 that flexible cable must be used. In flexible cables, the conductor is made up of very fine strands of conductive material, which enable the wire or cable to be bent and moved without causing damage. Eventually these strands do break though, but because there are so many, the effective current-carrying capacity is only reduced marginally.

ASSESSMENT GUIDANCE

The smallest size flex for lighting circuits is 0.5 mm^2, which has a current rating of around 3 A. This can be connected to a lighting circuit protected by a 6 A circuit breaker as the flex cannot be overloaded by the lamp.

Thermosetting cables

Many different types of thermosetting cable are used within electrical installation work. BS 7671 groups these together and gives a maximum conductor operating temperature of 90 °C in regulation 523.1 for conductors under normal conditions. This means that the insulation can withstand a higher temperature before it is damaged. This higher temperature requirement is imposed so that, under fault conditions such as short circuit, earth fault or even overload, the increase of current will still not damage the insulation, provided the fault is cleared quickly.

The conductor's normal operating temperature means that it can only carry a certain amount of current continuously; in Appendix 4 of BS 7671, tables 4E1 to 4E4 give the current carrying capacities of these conductors, taking this into account.

Constructional aspects of thermosetting cables

Thermosetting cables are so-called because of the material that is used as the insulation. The insulation is made out of a polymer resin that has been cured to form a plastic or a rubber. The curing process can be done in many different ways but the effect is the same in that the strings of polymer resin become linked to make a stronger material that is less likely to melt.

Two main types of thermosetting materials are used in cables as insulation.

Thermosetting insulated wire

- Cross-linked polyethylene (XLPE) has different polyethylene chains linked together ('cross-linking'), which helps prevent the polymer from melting or separating at elevated temperatures. Therefore XLPE is useful for higher temperature applications. XLPE has good ageing characteristics and resistance to water. Normal operating temperatures are typically between 90 °C and 110 °C, with a maximum temperature limit of 250 °C.

- Ethylene propylene rubber (EPR) is a copolymer of ethylene and propylene, commonly called an 'elastomer'. EPR is more flexible than XLPE, but has higher dielectric losses. Normal operating temperatures are typically between 90 °C and 110 °C, with a maximum temperature limit of 250 °C.

Thermosetting cables can suffer from 'water-trees', which reduces the insulating properties of the insulation. These are tree-like defects, filled with water, which develop in the insulation of cables. The defects usually originate from defects, voids or contaminants in the polymer, which occur during the cable manufacture. Water-trees occur only in the presence of water in the insulation and they are usually invisible to the naked eye in the dry condition.

size, based on the key features that determine the size of the cable. These are:

- number of conductors within the cable
- the gauge of the copper sheath
- the csa of the conductors within the cable.

Each accessory, and cable used would be identified, using the coding system. An example of a code might be:

2L1.5

This would mean that the cable was:

- two-core cable
- light-gauge copper sheath
- conductor csa of 1.5 mm^2.

Any pot assembly used would need to bear the same code as the cable. The code is stamped on the pot to ensure the correct selection is made. If the sheath were heavy-gauge, as opposed to light-gauge, then the code would change to 2H1.5.

Advantages and disadvantages of mineral-insulated cables

EMI

Electromagnetic Interference (EMI) is caused by some a.c. circuits, due to the current flow and resulting electromagnetic effect. EMI can cause surges and spikes, which can have an impact on circuits where the conductors are run alongside other circuits.

Cable	Advantages	Disadvantages
Mineral-insulated copper-clad (MICC)	<ul><li>High mechanical strength, able to withstand some impact.</li><li>Can work in extreme temperatures without damage to the insulation.</li><li>Very low dielectric losses.</li><li>Reduces the need to fit RCDs when used in prescribed zones.</li><li>Can be used for circuits that are sensitive to **EMI**.</li></ul>	<ul><li>Very expensive to purchase and install.</li><li>Needs special tooling to terminate.</li><li>Insulation becomes damaged if air gets into it.</li><li>Cable can crack when used in installations subjected to high levels of vibration.</li></ul>

One important point to remember when selecting mineral-insulated cables is that, as the sheath is made from copper and this has a very good connection to a brass pot at either end, the copper sheath can be used as the circuit protective conductor. Even if the sheath is not being used as the circuit protective conductor, then it must still be connected to earth to reduce the electromagnetic effects that the conductors will have on the sheath when the cable is carrying current.

The fact that the conductors are enclosed in a metal sheath means that the conductors are less likely to be affected by external electromagnetic influences, which can be a source of voltage disturbances.

ASSESSMENT GUIDANCE

A typical vibrating load would be a motor. The loop in the MICC cable would also allow movement of the motor if adjustment or alignment was needed.

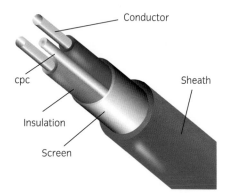

Conductor

cpc

Sheath

Insulation

Screen

FP200 fire-retardant cable

ACTIVITY

What other types of fire-resistant cable are there apart from FP? Find out from the internet or wholesalers' catalogues.

Fire-retardant cables

Mineral-insulated cables are excellent for use as fire-retardant cables. However, the cost of installation is often a very big drawback. Where the added benefit of mechanical protection offered by mineral-insulated cables is not required, alternatives exist that also offer a high level of resistance to the effects of fires. These cables are in many cases preferred, as they are quicker to install and cheaper to buy.

Typical fire-retardant cables will not only have a high temperature rating, they will also typically be halogen free, but they should not to be mistaken for screened or low smoke and fumes (LSF) cables.

Construction of fire-retardant cables

One of the more common fire-retardant cables used in electrical installations is referred to as FP200 cable, although this is not the only type.

The conductors are solid copper, surrounded by silicone-based insulation. There is also a non-insulated solid aluminium conductor among the insulated ones. The conductors are wrapped in a thin aluminium sheath, which is then covered with a final layer of PVC. The cable is very easy to strip, but this also presents a problem as any damage to the sheath can result in the sheath splitting and exposing the insulation.

Advantages and disadvantages of fire retardant cables

Cable	Advantages	Disadvantages
Fire retardant (FP200)	■ Cheap and easy to install.	■ Sheath can be easily damaged.
	■ Aluminium conductor can be used as CPC.	■ Low mechanical strength.
		■ Thin aluminium sheath cannot be used for EMI protection as it is not earthed.

ASSESSMENT GUIDANCE

FP is the identification used by one manufacturer. There are also FP 400 and FP600, which are armoured versions.

One important point to remember is that fire-retardant cable is not always an appropriate direct swap for mineral-insulated cable as there is little mechanical strength and it is also susceptible to EMI.

Armoured cables

Armoured cable is used where a high level of mechanical protection is required, but the use of conduit or trunking has been ruled out. One such application would be a cable buried in the ground. There are various forms of armoured cable. They use a mixture of different insulations and provide mechanical protection by a form of armouring. The way in which these cables are identified is normally by a mixture of the insulation and armour used. Some of the common terms used with regards to armoured cables include:

- PVC (polyvinyl chloride)
- PE (polyethylene)
- XLPE (cross-linked polyethylene)
- SWA (steel wire armour)
- AWA (aluminium wire armour)
- LSOH, LSZH (low smoke zero halogen)
- HOFR (heat, oil and flame-retardant).

The cables are identified by a mixture of the above abbreviations, based on the different layers of protection being used.

Constructional aspects of armoured cables

Due to the diversity of armoured cables it is not possible to describe every different type so the cables chosen are the most commonly used.

Cable	Picture	Construction
PVC/SWA/PVC		<ul><li>Copper or aluminium conductors</li><li>PVC insulation</li><li>PVC bedding</li><li>AWA armour for single core a.c. cables and SWA if multi-core a.c. cables</li><li>PVC sheath</li></ul>
XLPE/SWA/PVC		<ul><li>Copper or aluminium conductors</li><li>XLPE insulation</li><li>PVC bedding</li><li>AWA armour for single core a.c. cables and SWA if multi-core a.c. cables</li><li>PVC sheath</li></ul>

BS7671 Regulation 521.5.2 states

> 'Single-core cables armoured with steel wire or steel tape shall not be used for an a.c. circuit.'

The regulations go on to note that this is because the steel armour of a single-core conductor is regarded as a ferromagnetic enclosure. This means that it can be magnetised and if a single core conductor were to be used on an a.c. circuit it could induce a current into the cable resulting in overheating. This would not be detected by any overcurrent device and could lead to a risk of fire.

The constructional aspects of modern armoured cable differ from those of older styles but the basic principles are the same for all armoured cables. Each armoured cable has conductors for carrying current. These must be insulated and the cable will have some form of armouring to offer mechanical protection to the cable.

The construction of armoured cable is different from that of the other multi-core cables already identified. The conductors are stranded until they reach a cross-sectional area (csa) of 300 mm^2 or more; at a greater csa they are shaped solid conductors as they normally will not be installed with many bends. The conductors are each individually surrounded by insulation which is either made from thermoplastic or thermosetting material. There are, however, additional layers before the cable's sheath. The conductors are embedded in a layer of PVC called bedding. The bedding not only holds the cores together, but also provides mechanical protection from the armour for the insulation of the conductors.

Steel wire armoured cable

Surrounding the bedding is the armour, which is typically in the form of steel or aluminium wires but can be made of metal tape on the larger-sized cables. The wires travel along the cable in a slow spiral, so that the cable retains a certain amount of flexibility. On top of the armour is the sheath, which can be made from PVC or, where extra protection is required, from XLPE. The XLPE sheath does not split or tear as easily as PVC and it provides a higher resistance to damage from building materials and surfaces. When the cable is to be buried, the SWA is normally of XLPE/SWA/PVC construction.

ACTIVITY

What method of core identification do some armoured cables use as an alternative to colours?

Advantages and disadvantages of armoured cables

Cable	Advantages	Disadvantages
SWA	High mechanical strength	■ Needs a much larger bending radius than other cables. ■ Needs special glands to be installed. ■ Can take longer to terminate than other cables. ■ Single core versions cannot be used on a.c. circuits due to electromagnetic effects
AWA	High mechanical strength	■ Similar to SWA, but single-core versions can be used on a.c. circuits. ■ Can be a lighter option than SWA due to aluminium being used as the armour.

Though other forms of armoured cables are available, these tend to be the most commonly used. One other important disadvantage of armour over non-armoured cables is the size of the minimum **bend radius**.

The On-Site Guide, OSG, appendix D has guidance on the minimum internal bend radius of different types of cables and in the absence of manufacturers information, this can be used to identify how much space would need to be available to perform a bend in the cable.

Bend radius

This is a measurement of the tightness of the curve that can be applied when bending a cable. If the internal radius is too small then the cable will be subjected to undue stress and strain and may become damaged during use.

▼ **Table D5** Minimum internal radii of bends in cables for fixed wiring

Insulation	Finish	Overall diameter, d* (mm)	Factor to be applied to overall diameter of cable to determine minimum internal radius of bend
Thermosetting or thermoplastic (PVC) (circular, or circular stranded copper or aluminium conductors)	Non-armoured	d ≤ 10	3(2)†
		10 < d ≤ 25	4(3)†
		d > 25	6
	Armoured	Any	6
Thermosetting or thermoplastic (PVC) (solid aluminium or shaped copper conductors)	Armoured or non-armoured	Any	8
Mineral	Copper sheath with or without covering	Any	6‡

*   For flat cables the diameter refers to the major axis.
†   The value in brackets relates to single-core circular conductors of stranded construction installed in conduit, ducting or trunking.
‡   Mineral insulated cables may be bent to a radius not less than three times the cable diameter over the copper sheath, provided that the bend is not reworked, i.e. straightened and re-bent.

ASSESSMENT GUIDANCE

When installing cables on ladder rack or tray, it is the bending radius of the cable that determines the size of the ladder or tray used. If the ladder or tray is too small, the bend on the cable could be too tight.

Other types of armoured cable

In addition to those listed, an old type of armoured cabled that was used extensively in power distribution was paper-insulated lead-sheathed cable (PILC). These cables were predominantly used as buried cables in distribution networks and they have been identified as having a detrimental impact on the environment.

The impact is due to the lead sheath corroding and contaminating the surrounding area with high levels of lead. With rainfall, this sometimes resulted in raised levels of lead in the groundwater. Potentially, the water supply could be contaminated, leading to lead poisoning. New underground distribution network cables now tend to be XLPE sheathed.

Flexible control cables

In some electrical applications, cables are used for control purposes and interface wiring. Such cables may require additional mechanical and electromagnetic protection. These needs are met by a group of cables often referred to as control flex. They are denoted as CY, SY and YY types and they often come in standard multi-core arrangements, where the conductor insulation is black, but is printed to enable each core to be identified easily.

Constructional aspects of flexible control cables

As with all cables, flexible control cables have conductors, insulation and sheaths, but several factors make them different from other cables. The conductors of these cables are never solid, but stranded, due to the situations where they are used. They are common in applications that require movement of the cable while in use and are also used in locations subject to vibration.

The three different types of flexible control cable have different applications and so their construction differs slightly.

The number of conductors in flexible control cables ranges from two up to and including 80. Conductor sizes range from 0.5 mm^2 up to and including 120 mm^2, but not for all cable core numbers. Cables with larger numbers of cores only tend to be available with conductors up to and including 2.5 mm^2. Only three-, four- and five-core cables are available with the larger sized conductors.

ACTIVITY

Can CY, SY and YY cables also be used for supplying power to equipment?

Cable	Picture	Construction
CY		■ Flexible copper conductors ■ PVC insulation ■ Tinned copper wire braid screen ■ PVC sheath
SY		■ Flexible copper conductors ■ PVC insulation ■ PVC bedding ■ Galvanised steel wire braid screen ■ PVC sheath
YY		■ Flexible copper conductors ■ PVC insulation ■ PVC sheath

Advantages and disadvantages of flexible control cables

Cable	Advantages	Disadvantages
CY	■ Can be used in locations subject to vibration. ■ Can provide protection against EMI.	■ Requires special glands for termination.
SY	■ Can be used in locations subject to vibration. ■ Can provide protection against EMI.	■ Requires special glands for termination.
YY	■ Can be used in locations subject to vibration. ■ Cheap to install.	■ Has no protection from EMI.

Applications of flexible control cables

When using electrical equipment in an industrial or commercial environment, it is common for the wiring between parts of the installation and the equipment to involve more than just a simple supply of voltage and current. The wiring will also often require control signals and voltages to be connected between the installation and the equipment. All three of these flexible cables can be used to achieve this. Flexible control wires can all be used to connect between both fixed and mobile equipment and air-conditioning systems.

The main difference between the types of the cable is that the CY and SY cables are shielded or screened, whereas the YY has no shield or screen. A screened cable will prevent or reduce the impact of the EMI created by the circuit on surrounding circuits. A shielded cable will prevent or reduce external EMI from having an impact on the circuit it carries.

Both CY and SY cables may be used when the wiring is next to data cables and the equipment being controlled involves lots of signals, such as starting and stopping. Both of these types of cable come at additional cost, so if there is no risk of interference on either the control circuit or others, then the cheaper and easier option of YY would be used.

Data cables

Information technology, driven by the creation of the worldwide web, has brought about an increasing demand for data transmission. The infrastructure that supports our use of information technology has had to improve, in terms not only of capacity but also in reliability and speed.

Data cabling is often referred to as structured cabling. The earliest networks were simply two cables connecting computers together and were often very problematic. These networks have grown in size and complexity over time and the amount of data being passed between computers has also increased. To support this, the networks have had to evolve both to support the increase in demand and also to improve reliability.

To keep pace with the fast changes in this part of the electrotechnical industry, new cables have been developed and are constantly being improved. The way in which data cables are used remains the same, in that there must always be two conductors to send data. In some structured cable assemblies, a different pair of conductors is used to receive data.

Twisted pair

One type of data cable that is used within the industry is twisted pair. It comes in three main forms.

- Unshielded (UTP) is the most common form of data cabling, with eight cores twisted together in pairs. Twisting into pairs reduces the effects of eddy currents and EMI. Each pair has a different number of twists per meter, further reducing the effect of EMI within the cable.

- Shielded (STP) takes this protection further, with each pair covered by a foil. All the pairs are grouped together under another foil. The purpose of the foil is to ensure that the conductors are not affected by external or internal EMI. The foil must therefore be connected to earth at each end of the cable run.

- Screened (ScTP) is a combination of UTP and STP is. The cores are twisted together as in UTP and each pair is foil screened, but the difference is that the whole is enclosed in a braided screen.

The quality of the conductors and the number of twists, along with the size of the conductors, effects the quality of the signal at higher frequencies. The qualities are identified by the category (Cat) number allocated to them. Data cables are normally identified by these categories and it is important not to mix the different types on the same network. The same applies to the use of UTP, SCT and ScTP, as these must not be mixed either, or the protection provided by the likes of STP and ScTP will be lost.

All data cables are restricted by the length of run over which the signal can be sent. Resistance and interference cause the signal to degrade if it is sent too far. Most of the cable categories identified can only achieve their maximum output over a distance of up to 50 m. If the network cable is to be run over a distance greater than 100 m, then signal repeaters have to be installed to boost the signal.

The pairs of colours of the conductors are normally:

- brown with brown and white
- orange with orange and white
- green with green and white
- blue with blue and white.

The purpose of the colours is to show which pair is which. Remember that these cables should not be untwisted any more than they have to be, as this will then leave the conductors susceptible to EMI.

Twisted-pair cables: UTP and STP

ACTIVITY

What termination method is often used with UTP cables?

The categories of data cable

Category	Bandwidth	Applications
Cat 1	1 MHz	Was used for analog voice telephone systems. This was an informal category and was more commonly known as grade 1.
Cat 2	4 MHz	Used for old token ring networks for IBM systems and was typically a 100Ω UTP cable and could handle 1 Mbps. This was commonly known as grade 2.
Cat 3	16 MHz	Used for telephone and 10baseT networks. This was the standard for ethernet networks up the late 1980s. These cables could handle up to 10 Mbps.
Cat 4	20 MHz	Used for a short period again for token ring networks but this time with improved data handling of up to 16 Mbps.
Cat 5	100 MHz	Used for 100baseT networks for both voice and data applications, it can handle up to 100Mbps and was established as the minimum requirement for high speed data transfer.
Cat 5e	350 MHz	Used for 1000baseT networks for both voice and data applications, it can handle up to 1000 Mbps and quickly took over from Cat 5 cables as the enhanced standard requirement.
Cat 6	250 MHz	Used for 10GbaseT networks for both voice and data applications, it can handle data rates of 2.5 Gbps. (These cables are physically larger than the Cat 5 and Cat 5e cables.)
Cat 6e	750 MHz	Used for 10GbaseT networks for both voice and data applications, it can handle data rates of up to 10 Gbps.
Cat 7	600 MHz	Used for 10GbaseT networks for data and full-motion video.
Cat 7a	1000 MHz	Used for 40GbaseT networks intended to reduce data loss over distances greater than 50 m.
Cat 8	1200 MHz	Still under development.

Coaxial cable

This cable is expensive to run, but it can handle large amounts of data. The cable is similar to the coaxial cable that you would typically see being used for television aerials and satellite television cables.

Coaxial cable is made up of a solid copper conductor, which is then covered with PVC or Teflon insulation that acts as a packer to give the cable a structure. Then a copper wire mesh is woven around this to form a shield, which is then covered by a PVC sheath called a cable jacket.

■ *Through box* – this box has two spouts coming out of opposite sides. This type is often used as a through link and can be used over longer runs to make drawing cables easier.

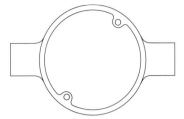

Through box

■ *Angle box* – this box has two spouts at 90° to each other. This is useful for going round corners or changing direction.

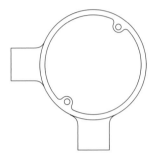

Angle box

■ *Tee box* – as its name suggests, this box has three spouts, forming a T-shape. It has the benefits of both an angle and through box.

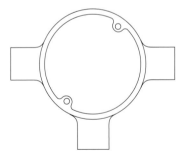

Tee box

■ *Four-way box* – this is sometimes called a cross box. It has four spouts at 90° to one another.

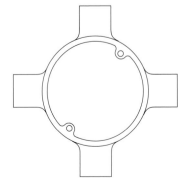

Four-way box

■ *Tangent-through box* – similar to the standard through box. This box has the spouts at one edge. This allows the box to be placed closer to a wall joint or ceiling.

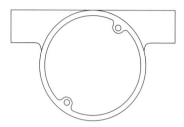

Tangent-through box

■ *Tangent-angle box* – similar to the angle box, but with the spouts positioned towards one side of the box.

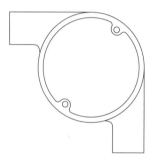

Tangent angle box

■ *Tangent-tee box* – this box has the same shape as the tee box, but due to the opposing spouts being on one edge, the conduit can be positioned closer to a ceiling or other edge.

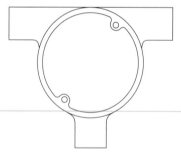

Tangent-tee box

■ *H-box* – this box has four spouts forming a letter H, with two on either side. It is similar to two tangent-through boxes joined together.

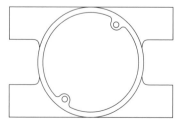

H-box

- *U-box* – this box allows for a 180° change of direction, by having two spouts on the same side of the box.

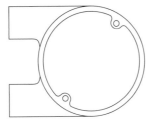

U-box

- *Y-box* – similar to the U-box, the Y-box allows for a 180° change of direction and has the benefits of a through box. The through spout is centrally located on the opposite side from the other two spouts.

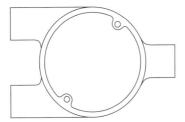

Y-box

- *Back-entry box* – this type of box is similar to a junction box with the spout or spouts located in the back of the box section. It can house up to four spouts.

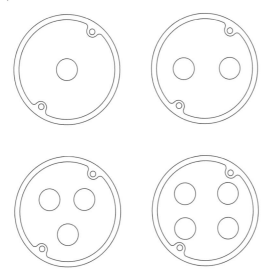

Back entry boxes

ASSESSMENT GUIDANCE

You can of course drill the back of any conduit box to form, for example, a back outlet H-box. There would not be a threaded spout, so termination would be by coupler and male bush.

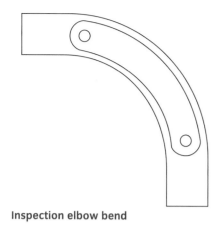

Inspection elbow bend

Conduit bends and elbows

You will learn to fabricate your own bends during assembly. However, pre-manufactured bends and elbows are also available and they tend to be one of two types:

- inspection joints
- non-inspection joints.

Inspection joints are usually preferred as these allow the cables to be drawn in easily during installation. However, if the bend is on a short run and within 0.5 m of the accessory, a non-inspection bend can be used.

Whatever type of bend you use, bear in mind compliance with BS 7671 Regulation 522.8.3, which states:

'The radius of every bend in a wiring system shall be such that conductors or cables do not suffer damage and terminations are not stressed.'

and regulation 522.8.6 which states:

'A wiring system intended for the drawing in or out of conductors or cables shall have adequate means of access to allow this operation.'

So, if inspection joints are not used, there must be adequate alternative access points along the conduit run. These can be provided by accessory boxes, such as through boxes on long runs.

Joining components

As mentioned earlier, conduit is available in standard lengths of 3 m and 3.75 m. Conduit runs may need to be joined. The basic method of joining two runs of conduit together is by use of a coupler.

The coupler that is used on PVC conduit is a piece of tube that has a larger diameter than that of the conduit. The conduit is simply slotted into either side of the coupler and secured in place with adhesive. Note that, under COSHH Regulations, the adhesive that is used in the joining of PVC conduit must be recorded on the National Poison Centre records; this should be done by the supplier or wholesaler selling the adhesive.

Expansion coupler

In a long run, PVC must be able to expand and contract. When the run of PVC conduit is 5 m or longer, an expansion coupler must be used.

This is similar to a normal PVC conduit coupler but is twice the length and adhesive is only used on one side, allowing the other side to move in and out of the coupler. It is important to ensure the level of penetration is correct, allowing for contraction without leaving the coupler and expansion without buckling.

Steel conduit also uses couplers but the use of adhesive is not permitted. All joints in steel conduit must be tight and electrically sound, so that if one part did become live, the installation would be connected to earth and the protective devices would operate.

To achieve this, the coupler is slightly larger than the conduit and there is a thread all the way through the inside of the coupler. This means that, once a thread is cut on both pieces of the conduit to be joined, the components can be screwed together.

In some joints it is necessary to have the thread on the outside. This is called a nipple. The nipple has the same size external diameter as the conduit and the thread is cut along its length. It is normally about the same length as a coupler and can be connected with a coupler.

Coupler and nipple

Erection components

There are various different ways of securing conduit to walls. If the conduit is to be covered by plaster, it is common to use a crampet to secure the conduit temporarily in place. This looks like a bent-over floorboard nail.

When the conduit is to be installed on the surface, it is common to use a type of *saddle* to hold it in place. All of the saddles shown here are available for PVC and steel conduits.

■ *Strap saddle* – this is used when the conduit is to be secured in place directly on the surface, with no gap at the back. It is also referred to as a stamp saddle.

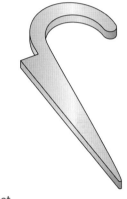

Crampet

Strap saddle

ACTIVITY

What is the advantage of using spacer bar saddles, over strap saddles, when mounting conduit?

■ *Spacer bar saddle* – this is the most common form of saddle and is available for both steel and PVC conduit. It allows the conduit to be positioned slightly away from the surface of the wall. The holes on the stamp are designed so that, when the saddle is mounted horizontally, the stamp will drop into place to make it easier for installation.

■ *Distance saddle* – this is similar to the spacer bar saddle, but with a thicker back plate. It can be used to install conduit on surfaces that undulate.

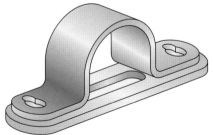

Spacer bar saddle

Distance saddle

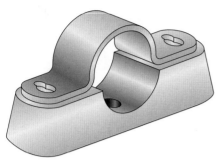

Hospital saddle

■ *Hospital saddle* – this is used where it is important to be able to clean behind the conduit. The hospital saddle has no horizontal edges for dirt and dust to accumulate and also moves the conduit away from the wall, just far enough to make cleaning possible. As its name suggests, these are used in some areas of hospitals, but they are also suitable for use in commercial kitchens along with stainless steel conduit.

The saddles used for PVC conduit vary slightly from that used for steel. Some forms of saddle do not require the use of screws to secure the stamp to the back plate.

Bushes

There are two types of bush for steel conduit, both of which are made of brass.

■ *Male bush* – the thread is on the outside and it is used to thread into a coupler or accessory.

■ *Female bush* – the thread is on the inside and is used where the thread on the conduit protrudes into an accessory and the cable needs to be protected from the rough edges.

For PVC conduit, the bushes are combined with the adaptors for securing the conduit. There are two types of bush and adaptor combination.

■ *Male adaptor* – this uses a part similar to the brass male bush used on steel conduit assemblies and screws into the adaptor.

■ *Female adaptor* – this uses a part similar to a lock ring and screws onto a thread already on the adaptor.

Sizing of conduit

When you are installing conduit, consider the fact that the cables will get *hot* and need to be able to cool to prevent overheating. There must be sufficient air space around the cables.

Conductors can only occupy 40% of the capacity of the conduit. The maximum fill capacity for conduit allows for airflow around the conductors.

Appendix E of the On-site Guide has standard tables giving the number of conductors that can be installed in a piece of conduit, taking into account all the relevant factors. The tables take into account the size and type of conductors, the length of the conduit run and how many bends there are in the run. Further information can be found in IET Guidance Note 1.

To use Appendix E, follow this simple and logical approach.

- Identify all the different sizes of wiring conductors used.
- Identify the quantities of each different size.
- Using the tables in Appendix E, select the correct cable factor for each size of cable.
- Multiply the factor by the number of conductors of that size.
- Add the totals together.
- Select the conduit with the next factor up from the total value.

An easy way to follow these steps is to use a simple table, filling in the data as you go. The example below is a conduit sizing calculation for three lighting circuits and one ring final circuit, wired using solid copper thermoplastic cables in a run of 2.5 m. The steel conduit is to be used as the protective conductor for each circuit.

Conductor size (mm²)	Number of conductors	Cable factor	Total cable factor
1.5	3 × 2 = 6	27	6 × 27 = 162
2.5	2 × 2 = 4	39	4 × 39 = 156
		Total	162 + 156 = 318

Example of a conduit sizing calculation

The calculation gives an overall conduit factor of 318. The appropriate table in Appendix E of the OSG, shows that 20 mm conduit can be used as it has a spacing factor in excess of the 318 needed.

ACTIVITY

Give reasons why 16 mm conduit is now rarely used.

▼ **Table E2** Conduit factors for use in short straight runs, taken from the IET On-Site Guide

Conduit diameter (mm)	Conduit factor
16	290
20	460
25	800
32	1400
38	1900
50	3500
63	5600

If – after checking BS 7671 Regulation 543.1.1 – it was found that 20 mm conduit did not have sufficient area to act as the protective conductor, then the calculation would have to be repeated allowing for additional circuit protective conductors for each circuit.

Trunking systems

SmartScreen Unit 305

Handout 18

As with conduit systems, there are various forms of trunking system. Trunking is the generic name given to the enclosure of a wiring system that has a rectangular or box-section fabrication and a removable lid. The purpose of trunking is to enable wiring to be installed easily and to provide the wiring with a level of mechanical protection.

Unlike conduit systems, trunking systems only come in one of two forms:

- PVC
- steel.

There are benefits that trunking can provide which are not available when using conduit. Trunking can be used to install cables and wires alike, and in the same run. Trunking can have wiring added in at a later point, without the need to remove existing wiring.

PVC trunking

PVC trunking comes in various arrangements.

- Dado trunking can be split into sections to enable power and data cables to be run within the same trunking arrangement and is common in offices and educational establishments. It is commonly referred to as segregated trunking.
- 'D-line' profiled trunking can be used to blend into the existing wall furniture and can be concealed as skirting board or even in the corners of rooms without the need to chase out walls.
- Mini-trunking is commonly used where only a few wires are required to be run and the likelihood of additional cables at a later point is unlikely. Mini-trunking typically will have a snap on lid. It can often be mounted by means of a self-adhesive strip on the back.

Two of the main benefits of PVC trunking over steel is that it cannot have an emf induced within it and it can be made in a variety of different colours. As PVC trunking can blend into its surroundings, it is useful in domestic and office environments where steel would be obtrusive.

Segregated trunking

Irrespective of whether the trunking is steel or PVC, segregated trunking can be used to contain wiring for different voltage bands and different types of circuit, as it can be split into sections. This is called segregation. It is important to remove potential interference from sensitive circuits and circuits of different voltage bands. The use of different compartments or sections can assist with segregation of specific circuits; for example, data cables can be segregated from low-voltage power circuits, and band 1 and band 2 circuits can be separated.

Segregated trunking can even have separate lids for each section, so that only the lid of the partition being worked on is opened. This is particularly useful in rooms where computers are positioned around the perimeter of the room. This type of trunking, when installed on the surface, is commonly referred to as dado trunking.

Files are not just described by their shape or purpose. They are also identified by coarseness – some files remove a lot of metal but leave a rough finish and others remove less but leave a very smooth finish.

The coarseness is identified as a grade of cut and the levels include:

- very smooth (finishing cut)
- smooth (dressing cut)
- second cut
- bastard
- rough.

As with any hand tool, it is important to make sure your cutting tool is safe to use. So, all files should be fitted with a handle which is secure and the faces of the file should be kept clean with a stiff wire brush. Damaged or blocked teeth on the file will result in an uneven cut, which may cause slippages that result in injury to the user.

Wire strippers

Wire strippers provide a safe and reliable method of removing the insulation from a wire or cable, without damaging the conductor. Wire strippers come in various designs, but the common principle is that the cutting jaws only cut into the insulating material and not into the conductor.

Automatic wire strippers

Wire strippers can be either manual set (by a screw or dial) or automatic set. Side cutters can also be used but these often damage the conductor.

Cable cutters

There are a variety of different tools that can be used to cut cables, the most common being side cutters. A side cutter is probably one of the most important tools that the electrician has in the tool kit. They can be used, for example, for cutting cables to length, cutting sleeving and cutting nylon tie-wraps. They work on a compression-force basis and are shaped so that the cutting edge is along one side.

Depending on the size of the cutters, cable sizes up to 16 mm^2 conductors can be cut easily. Larger cables, however, require larger types of cutters. These range from cable loppers up to hydraulic manual pump cutters for cables up to 300 mm^2.

Side cutters
Courtesy of Axminster Tool Centre Ltd

Screwdrivers

An electrician will use a selection of different sized screwdrivers. They will all have one thing in common in that they will all be of insulated construction to **VDE** standards.

The most common screwdrivers used include:

- terminal (3–3.5 mm)
- large flat (4–5 mm)
- pozi drive (PZ2)
- and, more recently, a consumer unit screwdriver.

VDE

An acronym of *Verband der Elektrotechnik* (originally the Association of German Electrical Engineers and now the Association for Electrical, Electronic and Information Technologies, in Germany) which is responsible for testing and certifying tools and appliances.

Electrician's screwdrivers

Electrician's knife

The electrician's knife is available in various arrangements but usually has a folding blade. The most common use of the electrician's knife is for stripping the outer **sheaths** of some cables. The blade generally has a shaped section to aid the removal of certain cable sheaths, such as on armoured or mineral-insulated cables.

Electrician's knives
Courtesy of Axminster Tool Centre Ltd

An electrician's knife should never be used for stripping the insulation from cables or for cutting cables to length. It should always be used pointing away from the body.

Specialist tools

Certain activities require tools that are specifically designed for the task and are only used for that task. For example, when working with steel conduit, it is necessary to use stocks and dies, along with reamers and engineer's squares. These and other tools are explained later in the unit.

SmartScreen Unit 305
PowerPoint Presentation 3

Sheath

The sheath of a cable is the outer covering which not only holds the cable together, but also provides a basic level of mechanical protection against light damage that may be caused during installation or use.

KEY POINT

It is dangerous to use a disposable knife blade when stripping cables, as the blade can break, creating sharp shards.

Die stock

Combination square

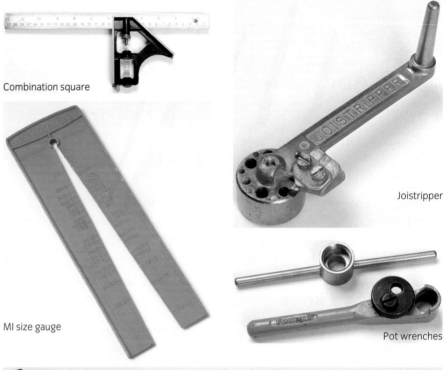

Joistripper

Ringing tool

MI size gauge

Pot wrenches

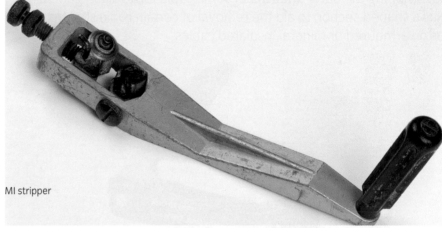

MI stripper

MI roller

ACTIVITY

What method or tool, is used for bending PVC conduit?

Conduit bender

A selection of electrical tools

Wrenches

A wrench is used to tighten nuts and/or bolts. Wrenches are often used in place of spanners, as one tool can be used on a variety of different sizes of nuts and bolts.

Some typical wrenches used in our industry are described here.

■ *Pipe wrenches* have jaws that are designed to fit around a selection of different nuts and bolts, and so they are ideal for tightening lock nuts and cable glands.

■ *Footprints* are used for some larger cables that require glands that are in excess of 60 mm in diameter and are often too big for a pipe wrench. Footprints can be set to large sizes and are well suited for tightening these larger glands and nuts.

| Pipe wrench or Stillsons | Water pump pliers | Adjustable spanner | Adjustable spanner | Combination spanner |

A selection of wrenches

■ *Bush spanners* are designed to fit into small areas. When installing metal conduit, it is important to ensure that all connections are tight. Some of the assembly components are called bushes. These are often small and in positions that are not accessible to pipe wrenches and footprints. A bush spanner enables you to tighten the bush correctly.

It is important to select the correct wrench for the job and to set it to the correct size. Incorrect use can damage the nut or bolt being tightened.

Bush spanner

ASSESSMENT GUIDANCE

Using the correct size of spanner is better than using grips, as they can slip and chew off the edge of the nut.

Hammers

There are several different types of hammer that can be used in our industry including:

- *Claw hammer* – this is the electrician's general purpose hammer. It is used for driving and removing nails and tacks. It typically weighs between 16 and 28 ounces (450–800 g).

- *Ball peen (or pein) hammer* – this hammer is typically used when bending and forming metal. It weighs between 4 and 32 ounces (113–800 g).

- *Engineer's hammer* – sometimes referred to as a baby sledgehammer, this can be used during light demolition work. It typically weighs between one and five pounds (450–2300 g).

Claw hammer

Ball peen hammer

Engineer's hammer

A selection of hammers

The most common mistake that people make when using a hammer is to grip it at the wrong place. The handle is designed to be gripped at a point on the handle furthest away from the head of the hammer, enabling the full force to be applied.

If a hammer has a split or damaged handle, or shows signs of damage to the head, it should not be used – it should be replaced immediately. Never hit the heads of two hammers together as this may result in parts breaking off.

BASIC SAFETY

When using any hand tool it is important to ensure that you take basic safety requirements into account. Ask yourself these questions.

- Is the tool in good condition?
- Is the tool sharp?
- Is the tool the correct one for the job?
- Do I know how to use the tool?
- Is the tool safe to be used?

The key point when using any tool is to check to see if it is safe to be used. A cold chisel that has started to bend over where it is hit with the hammer is displaying 'mushrooming'. The end of the chisel becomes deformed and bends over, like the cap of a mushroom. These bent-over pieces can break off, cutting your hand or, even worse, ending up in your eye.

Blunt tools require more force to perform cutting functions – which can lead to dangerous slips. If a saw blade is blunt, you may be tempted to apply more pressure, which could shatter the blade. The blade may also become hot with the potential to burn you.

Using a screwdriver as a chisel or lever is not a correct use of the tool. The shaft of the screwdriver may snap or slip. You could damage the handle, so that next time you use the screwdriver it ends up sticking into your hand.

Buying cheap tools can be a false economy; cheap tools tend to break easily, whereas good-quality tools often last a lifetime if looked after and used properly.

POWER TOOLS

Though most tasks can be performed using hand tools, the work is often performed quicker with **power tools**.

Power tools vary in size, from hand-held to larger machines that are mounted on a workbench or bolted to the ground.

Safety checks for power tools

When using power tools, it is important to make additional safety checks to those already mentioned for hand tools. The use of electrical energy adds extra risk. Here are some additional questions to ask.

- Is the item of equipment correct for the job?
- Is the casing of the tool without damage?
- Is the flex of the tool undamaged (if fitted)?

Assessment criteria

5.1 State the procedures for selecting and safely using appropriate hand tools, power tools and adhesives for electrical installation work

5.3 State the criteria for selecting and safely using tools and equipment for fixing and installing wiring systems, associated equipment and enclosures

5.4 State the criteria for selecting and safely using fixing devices for wiring systems, associated equipment and enclosures

ACTIVITY

State the possible detrimental effect of:

a) using a blunt masonry drill

b) using a 10 mm HSS to drill into a steel block, without first creating a pilot hole

c) using a 12 mm HSS to drill into 2 mm sheet steel

d) applying excess pressure to a 20 mm flat bit.

KEY POINT

Any tool is only as safe as the person using it. Using the wrong tool for the job increases the risk of injury and damage.

Assessment criteria

5.1 State the procedures for selecting and safely using appropriate hand tools, power tools and adhesives for electrical installation work

5.3 State the criteria for selecting and safely using tools and equipment for fixing and installing wiring systems, associated equipment and enclosures

Power tool

Any tool that requires electrical energy to make it work. The electrical energy can be obtained from batteries or from a mains supply.

Chuck

The part of the drill used for holding the drill bit.

Bit

The part of the drill that does the cutting.

Swarf

Chips and spirals of waste metal produced when cutting metal.

- Is the plug undamaged (if fitted)?
- Is the start/stop button in good condition?
- Are all guards in place (where fitted)?
- Is there a means of stopping the tool in case of emergency?

If you answer 'no' to any of these questions, the tool should not be used and the person responsible, such as your supervisor, should be notified immediately.

Drills

There are several different types of drill that can be used in our industry, for different tasks. It is important to recognise the best tool for the job and to ensure it is safe to use.

All drills have some common parts, such as a **chuck** and a **bit** (with some requiring a special tool called a chuck key to tighten and loosen the jaws of the chuck). In the past, the chuck key was often lost or misplaced, so some modern drills have a chuck that can be tightened by hand without the use of a special tool or key. It is important to make sure that, when securing a drill bit in the chuck, it is correctly positioned, straight and central, so that it does not wobble or come loose.

Electric hand drill

This is the most common type of drill and can be mains powered from either 230 V or 110 V a.c., as well as being available in battery form. This drill is probably the most versatile as it is a general-purpose tool.

The hand drill, as its name suggests, is portable and can be used in most positions. Consideration should be given to dust and **swarf** that may be generated during the drilling process – make sure suitable precautions are taken to minimise risk of injury.

Electric hand drills

Right-angled electric drill

The right-angled electric drill is a variation of the electric hand drill. This tool is often used when making alterations to an existing property and new holes need to be cut through the joists. The spacing of the joists often prevents access by a regular hand drill and so a right-angled drill is required.

The chuck is positioned at right angles to the rest of the body of the drill, so the space required for access is only that of the chuck and drill bit. Generally, this is less than the spacing of the joists.

Using a right-angled drill

Hammer drill

This type of drill is for drilling into masonry walls, floors or ceilings. The drill has a hammer action, which makes cutting more efficient. Some electric hand drills have this option built in and it can be easily selected.

When in hammer mode, the chuck is subjected to a set level of vibration that is transmitted to the drill bit. This helps the drill knock the bit into the masonry. This function must not be used when drilling through timber or steel, as it will damage the drill bit as well as the material. Masonry drill bits (for use in hammer mode) are designed and built to withstand these stresses.

Using a hammer drill

ASSESSMENT GUIDANCE

A variation of the hammer drill is the SDS drill which is particularly suitable for drilling into concrete and other hard structural materials.

Pillar drill
Courtesy of Axminster Tool Centre Ltd

Pillar drill

Where drilling needs to be very accurate and consistent, a pillar drill is often used. The pillar drill has an adjustable table, which can be positioned at various heights. The table is used to clamp the material to be drilled so that it does not move, so that a hole can be drilled in a precise location.

Drill bits

The term 'drill bit' refers to the item that is positioned in the chuck to perform the cutting process. Bits come in various shapes and sizes, according to the material to be cut. When drill bits are used for cutting the wrong material they can become blunt, very hot and damaged.

When a drill bit becomes blunt, it is tempting to press down more on the drill to make the bit cut. This increases the risk of injury as the bit could shatter. In addition, a blunt drill bit can become so hot that it melts, further increasing the risk of injury.

All drill bits are designed to cut in a clockwise direction. If a reversible drill is used, care should be taken to ensure correct rotation. A drill bit should never be run in reverse as this will damage it.

Brad-point bit

Brad-point drill bits are specifically designed for drilling into wood and have a cutting face that includes a centre spike. This spike cuts a pilot hole for the drill to follow, making sure the drill bit stays on track. Without this centre spike, the drill bit could go off track due to the grain of the wood.

ASSESSMENT GUIDANCE

Most drill bits sold in tool stores have a cutting angle of 118°, which is suitable for most applications. To cut this angle by eye is very difficult and special guides are available. Trying to sharpen small drills is probably not economically viable as a full set of drills, from 1 mm to 19 mm, can be purchased for a few pounds.

KEY POINT

Note the special design of the brad-point drill bit that feeds cut wood away from the drill bit point.

The flutes (which allow the cut wood to be fed out of the hole) are also of a wide design – the debris must be removed quickly from the hot tip or it may catch fire.

Hole saw

As the name suggests, this is a saw bit for cutting holes. There are two types: one for wood and the other for steel. Both types fit onto an arbour that holds the saw in place and has a pilot drill to make sure the saw acts around a central point.

A hole saw is ideal for cutting holes in sheet material or housings where the material is relatively thin and the hole size is larger than that of a conventional drill bit. Some of the more common saw sizes are 16 mm, 20 mm, 25 mm, 32 mm and 60 mm diameter.

A brad-point drill bit

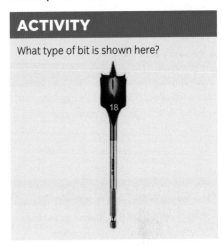

ACTIVITY

What type of bit is shown here?

Hole saws

Metal drill bit

Metal drill bits are sometimes referred to as **HSS** bits. These are intended to be strong enough to cut through steel and often have extra coatings, such as metal carbide, to make them resistant to wear.

The cutting face on an HSS bit is straight and there are two parts. A sharp HSS bit should look like a numeral 8, when viewed from the cutting end. The cutting faces are set at an angle from the horizontal, and the back of the face should be lower than the front, to ensure only the cutting edge is in contact with the material being cut. If the back edge of the cutting face is too high, it will rub against the material, causing excess heat and stopping the cutting by the drill bit.

HSS bits can be used to cut alloys such as aluminium, but they need to be run at a different speed.

A HSS drill bit

KEY POINT

Note the design of the HSS drill bit, which minimises friction when cutting.

Masonry drill bit

A masonry drill bit

Use of an HSS bit on masonry will quickly blunt the bit, depending on the make of bit and the masonry being drilled. The flutes on an HSS bit are designed to move metal swarf, not masonry dust particles .

A masonry drill bit is designed with a larger cutting face diameter to that of the drill bit shaft. This enables the masonry dust to be pushed away from the cutting face and along the flutes.

The cutting face of a masonry bit is made from a hardened material, usually tungsten carbide, that is designed to cut into hard materials and to withstand high temperatures.

Core bit

A core cutter is used to produce a large hole in a masonry wall. It is similar to a hole saw but specifically for masonry.

Core drill bit

Electric screwdriver

Pressure to work quickly in installation meant electric screwdrivers were inevitable. They all have common characteristics:

- bi-directional for tightening and loosening
- adjustable torque setting to prevent over tightening
- variable speed, generally controlled by the start button.

Some electric screwdrivers can also be used as electric hand drills. However, not all electric drills can be used as screwdrivers and, if in doubt, no attempt should be made to use them in that way.

KEY POINT

Note the special design of the masonry drill bit, which pushes masonry dust away from the drill bit

ASSESSMENT GUIDANCE

Battery drills have the obvious advantage of no trailing lead, though a second charged battery should always be available.

ACTIVITY

Identify a risk when using an electric screwdriver to tighten terminals.

Electric screwdrivers

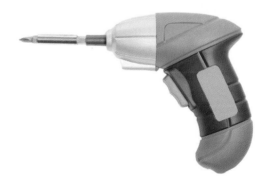

Personal protective equipment (PPE)

The use of power tools always presents a safety risk to the person using such tools, as well as those working around them. Consideration should always be given to the use of additional PPE.

When drilling, you should think about the damage that you may be doing to your ears, eyes, hands and lungs – always consider the use of ear protection, gloves and face masks. As a *minimum*, eye protection should be worn when using power tools.

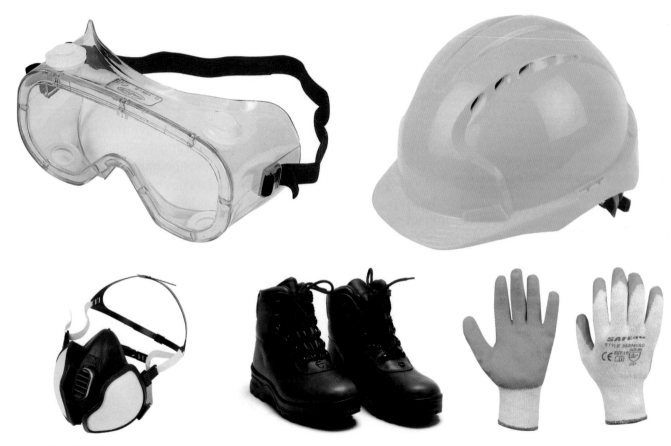

A selection of PPE

Before any drilling is undertaken, always check to ensure that you are not going to drill into something behind the surface you are working on. Always be alert when drilling, as hidden dangers often lurk, such as buried cables, pipes or even asbestos.

Assessment criteria

5.1 State the procedures for selecting and safely using appropriate hand tools, power tools and adhesives for electrical installation work

5.3 State the criteria for selecting and safely using tools and equipment for fixing and installing wiring systems, associated equipment and enclosures

ASSESSMENT GUIDANCE

The use of too much solvent or adhesive can block the tubes being joined.

KEY POINT

Unconsciousness and even death can result from exposure to very high concentrations of solvent vapours.

SOLVENTS AND ADHESIVES

In our industry we use a wide selection of solvents and adhesives, but not many people understand the difference. Chemical substances that are used to dissolve or dilute other substances and materials are called 'solvents'. Industrial solvents are often mixtures of several individual substances. They can be found under a variety of trade names.

Adhesives are used to join two or more items together and most adhesives come in solvent form. When installing electrical equipment, it is frequently necessary to use adhesives during the erection process. One common instance of using a solvent type adhesive is during the erection of a PVC conduit containment system. It is important to understand how to select the right solvent adhesive and what precautions you may need to take, in order to keep both yourself and others safe.

Solvents are very common and they can be found not only in adhesives, but also in cleaning and degreasing materials, paint removers, paints, lacquers and varnishes, and many more products. The use of solvents can be widespread throughout the working environment and so it is important to understand their potential impact.

Hazards

The HSE definition of a hazard is 'something with the potential to cause harm': solvents certainly fall into this category. It is important to understand that no two solvents are the same and each different type brings different hazards and can affect your health in different ways. Some of the short-term effects include:

- irritation of eyes, lungs and skin
- headache
- nausea
- dizziness
- light-headedness.

When you are affected by solvents, there may be an increased chance of having an accident. There can also be long-term effects on your health from repeated exposure to particular solvents. These may include dermatitis. Other impacts on health vary, depending upon the solvent to which you are exposed.

Exposure to solvents

The harmful effects of solvents are experienced due to solvents entering the body in various different ways. Solvents, like most liquids, give off vapours or fumes that can be inhaled. When they are being used, they can come in contact with your skin and some solvents can thus be absorbed into your blood system. The other most obvious method of solvents entering the body is through being swallowed. This may occur directly or because of a residue left behind, that is later transferred to an item of food or drink.

If a person feels their health is being affected by the solvents being used during their work, then their first action should be to talk to their supervisor and safety representative. If any of the effects of solvents, either the short-term or long-term, are present then the person should seek medical advice from a doctor as soon as possible.

ACTIVITY

Where expansion of PVC conduit must be allowed for, what type of sealant is used?

Precautions to be taken

There are some precautions that can be taken, whenever work that involves the use of solvents is to be undertaken. Before any person uses solvents, they should be trained in the storage, use, hazards, precautions and disposal of the solvent they are to use. Further precautions that can be taken by workers using solvents are listed here.

Preparation	<ul><li>Make sure that the risk assessment for the solvent has been read and can be followed.</li><li>Read the suppliers' safety data sheets and the container labels and follow the advice on them.</li><li>Seek advice if clarification of information is required.</li><li>Identify if alternative safer solvents are available to be used.</li><li>Ensure that the emergency actions are understood and can be performed in the case of an incident.</li></ul>
Control of vapours	<ul><li>Ensure that ventilation equipment is used when provided, to remove any vapours from the work environment.</li><li>Report any damaged or defective ventilation equipment so it can be replaced or repaired.</li><li>Wear respiratory equipment or protection that has been provided, no matter how silly it makes you feel.</li><li>Make the most of natural ventilation where possible and acceptable, by opening windows and doors.</li><li>Only use the minimum amount to complete the job correctly and replace the lid or cap on the container when the solvent is not in use.</li><li>Use sealed containers for solvent-contaminated waste materials, including any cloths and rags used.</li></ul>

Skin contact	■ Avoid skin contact at all times by wearing suitable protective clothing when using or handling solvent-based products.
	■ Do not used solvent-based products to clean skin, for example, do not use paint remover to remove paint from your skin.
Hygiene	■ Always wash your hands after you have used solvent-based products, especially before eating or drinking – or even smoking.
	■ Do not consume food or drink in an area where solvent-based products have just been used or are stored.
	■ Do not smoke near areas where solvent-based products are being used or are stored.
Confined spaces	■ Make sure breathing apparatus is used at all times.
	■ Make sure protective clothing and PPE is worn at all times.
	■ Ensure emergency evacuation procedures are in place and understood by all involved.
	■ Ensure that the company safe scheme of work is understood and seek advice if you have any questions.

Assessment criteria

5.4 State the criteria for selecting and safely using fixing devices for wiring systems, associated equipment and enclosures

SmartScreen Unit 305
Handout 29

FIXING DEVICES

If you have ever done any home DIY work you will no doubt be aware of lots of different ways in which items can be mounted on walls, ceilings and other surfaces. The reasons for the existence of so many different fixing devices are related to the ways in which buildings are constructed and the varying weights of what is to be mounted. Each fixing device serves a different purpose.

Screw-type fixings

A screw-type fitting is the most common type of fixing device. Screws can come in several different styles.

■ Round or pan-head fixings are used when it is necessary to mount metal enclosures to walls, without the edges of the screw head protruding, but the metal enclosure is too thin to have a recess formed to conceal the screw.

■ Countersunk head fixings are used when it is necessary to mount enclosures and other accessories and a recessed area can be formed within the enclosure. The internal taper of the head of the screw is designed to fit within the recess of the enclosure.

■ Self-tapping screws are used when a screw is being used to secure onto thin sheet metal and there is no existing thread enabling the screw to be inserted. The self-tapping screw will cut a thread as it is being screwed in.

Round head and countersunk screws

ACTIVITY

When attaching flat materials such as metal trunking to a wall, what type of screw head is recommended?

The definition of a screw is an externally threaded, headed fastener which is tightened by applying torque to the head, causing it to be threaded into the material it will hold. A screw is used to fasten something onto another item and the screw shaft will not penetrate both items. A screw will typically have a thread that will cover the entire length of the shaft. However, some wood screws only have a thread along part of their shaft.

KEY POINT

If the incorrect sized screwdriver is used then both the screwdriver and screw can become damaged.

Wall plugs

A screw is not designed to cut a thread into masonry. A liner must be used to provide a softer material for the screw to cut into. The liner will in turn grip the masonry wall. One such device is called a wall plug.

The wall plug is designed to fit into a pre-drilled hole and is split so that, as the screw threads into the wall plug, it expands the plug to fill the space and therefore exerts a strong force on the masonry with a strong resulting grip. Some wall plugs can support a mass of up to 20 kg, if installed correctly.

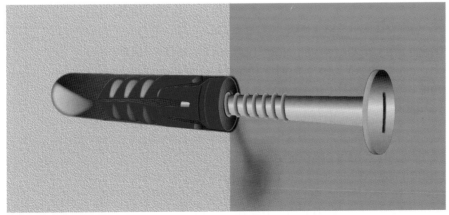

A wall plug in a wall

Bolt type fixings

Bolts differ from screws in that the shaft must pass through both items that are to be secured together and a bolt has to be used in conjunction with a nut. To obtain a reliable and repeatable fastening, the bolt–nut combination should always be tightened by holding the bolt head stationary and turning the nut. This type of fixing can come in various different forms.

Hexagonal-headed bolts 	This is the most common type of bolt. It has a hexagonal head and a hexagonal nut. A spanner or socket must be used to secure the head of the bolt, while a second spanner or socket is used to turn the nut to tighten it onto the bolt.
Coach bolt 	Unlike the hexagonal-headed bolt, this bolt is used when only one side can be reached at a time. The square section just below the head of the bolt bites into the material being bolted together as the nut is tightened and this stops the bolt from rotating as it is tightened. These are also sometimes referred to as carriage bolts.
U-bolt 	This bolt is bent into the shape of the letter U and has a thread at each end of the shaft. These bolts are generally used for securing pipes in position.
Eye-bolt 	When heavy equipment or assemblies need to be moved or lifted an eye-bolt can be fastened to the equipment or assembly and then ropes and chains can be secured through them to aid with the lifting.

ACTIVITY

What kind of fixing is shown here?

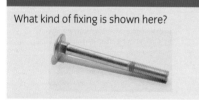

ASSESSMENT GUIDANCE

There are fixings available which screw directly into a predrilled pilot hole in concrete.

Ragbolts and rawlbolts

These are similar to wall plugs but are used for much heavier and physically larger items. Both ragbolts and rawlbolts need first to have a hole drilled into the masonry, but from then on the two fixings are installed differently.

A ragbolt is mainly used when the item to be secured in place is vertical, such as a post or mast. Once the hole has been drilled, the ragbolt is inserted into the hole, which is then filled with cement or concrete and left to set. The ragbolt is shaped so that when the cement or concrete has set, the bolt is held in place.

Use warm water to soften old, absorbent wall coverings, then use a wide wallpaper scraper to strip the paper off the wall. A steam-generating wallpaper stripper makes the job easier but it can cause damage to older walls. When removing painted wallpapers or washable wall coverings, scratch the surface with a wallpaper scorer to enable the moisture to penetrate. You can peel vinyl wall coverings off the wall, leaving the backing paper behind. If it is sound, just paper over the backing, or strip it like ordinary wallpaper.

Using the replacement wallpaper, measure the size of the drop required, taking care to match any pattern where possible. Cut the paper a little bit longer than required, to allow for trimming when the paper has been hung. Mix a small amount of wallpaper paste, according to the manufacturers' instructions, and then, using a broad brush, apply the paste sparingly to the back of the cut drop of paper. Make sure that all the back of the paper is covered.

Stripping wallpaper

Position the pasted wallpaper onto the wall and then, using a dry, soft brush, smooth the paper into position, making sure to brush out any air bubbles and remembering always to work outwards from the centre. When the drop is in place, trim the surplus, from the top and bottom of the piece, using a very sharp trimming knife and if necessary, a steel straight-edge.

The alternative to doing this yourself is, of course, to hire in a professional or agree an action with the client before the work is commenced. Some clients will use this as a reason for some redecorating.

> **KEY POINT**
>
> When pasting the paper, the paper can be folded back on itself to make it easier to carry.

> **KEY POINT**
>
> Do not get any of the paste on the front of the paper as this will mark the paper.

> **ACTIVITY**
>
> Do not use the customer's dining room table for pasting wallpaper. What kind of table should be used?

Pasting wallpaper

Replacing wall tiles

Wall tiles will be found in most installations near sinks, baths or showers. Once tiles become damaged, they cannot be repaired and need to be removed and replaced. This is a problem, as the damaged tile may be surrounded by other tiles that could also be damaged during removal of the original tile. Great care must be taken throughout the tile removal process.

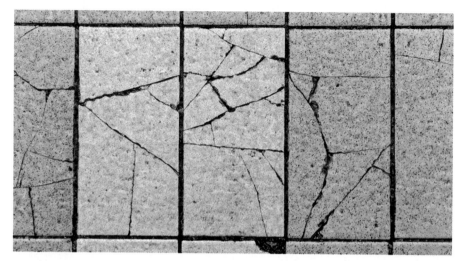

Damaged wall tiles

Before you can remove the damaged tile, you will need to weaken it by drilling 5 mm diameter holes in the surface. Most tiles have a smooth, glazed finish, so it can be helpful to use masking tape to create a large diagonal cross on the tile. Then, using a masonry drill bit, drill a series of holes at regular intervals along the two lengths of tape with holes not more than 10 mm apart.

It is important only to drill through the tile and not into the wall, so use a drill depth gauge. A replacement tile can be used to establish the depth of hole required.

Once you have weakened the tile by drilling the holes, use a grout rake to remove the grout from around the tile. Take care to ensure that the surrounding tiles are not damaged when using the grouting rake. Sometimes tile spacers will have been used to ensure equal spacing between tiles. These plastic spacers can often cause a grout rake to slip, so take particular care at each corner.

Tile spacers

With all the grout removed from around the damaged tile, the next stage is to start removing the tile. Using a sharp cold chisel and hammer, chip away at the tile, starting from the middle, and work out towards the edges along the masking tape. Care must be taken to avoid damaging the wall by hitting too hard, or through the chisel slipping.

Once the tile has been broken along the masking tape, remove the tape and clear away any of the tile that has come loose. The remaining pieces of broken tile also need to be removed. This is best performed using an old wood chisel, bevel-side down, underneath the remaining sections of tile.

With the tile now removed, it is important to prepare the wall for the replacement tile. Start by using the wood chisel to continue to remove the old adhesive. It is a good idea to use the new tile as a depth gauge to establish how much old adhesive needs to be removed, and from where. In order for the new tile to be easily and safely removed from the hole, it is a good idea to attach a piece of masking tape, as a tab, to the front of the tile to provide something to grip.

Once the wall has been cleaned to the correct depth, use an adhesive comb to apply tile adhesive to the back of the new tile. Having already used the tile to see if the wall was level will have given an indication of how much tile adhesive is needed. Use the tab to grip the tile and position the new tile in the prepared hole, making sure it is flush with the surrounding tiles.

The tile can be evenly spaced by using tile spacers, but this time with them positioned just in between the gaps at the bottom and sides rather than at the corners of each tile. With spacers in place, the tile adhesive must be left to harden. The time to set will vary, depending on the type of adhesive used, but typically, fast-setting adhesive requires about four hours to set.

Once the adhesive has set, the spacers can be removed and the space around the tile can be filled with grout to reseal the tiles. The grout can be applied with a grout-finishing tool and then any excess can be polished off with a dry cloth, once the grout has dried.

Fast-setting tile adhesive

Tidying up

Once any repair work has been performed, it is important to clean up properly. This includes removing all rubbish and waste materials. Dust sheets and covers that have been used must be carefully folded in on themselves so as not to generate any more dust and to capture any residue inside the sheets.

With the dustsheets and protective covers removed from the work area, it is time to use the most important tools in terms of customer relations, the dustpan and brush and vacuum cleaner. Ensuring the work area is clean and tidy after you have finished, will often go a long way towards bringing repeat business from many clients. It also provides a chance to confirm that all tools have been collected and packed away, rather than being lost or misplaced.

Know the regulatory requirements which apply to the installation of wiring systems, associated equipment and enclosures

Assessment criteria

7.1 Specify the main requirements in accordance with the current version of the IET wiring regulations and describe how they impact upon the installation of wiring systems, associated equipment and enclosures

KEY POINT

This outcome will focus on the content of BS 7671 and so it is essential that the reader has a copy of BS 7671 to refer to when studying this material.

INTRODUCTION TO BS 7671

Throughout this course, reference has been made to BS 7671. It is important to understand the value of this book. BS 7671 has been amended and updated several times during its existence and some people complain about having to purchase new copies of the book to keep up to date. While it is true that electrical principles such as Ohm's law do not change, BS 7671 is about putting the principles into practice and as our use and understanding of electricity keeps growing, then the guidelines we follow have to change as well.

Regulation 114.1 states:

'The regulations are non-statutory. They may, however be used in a court of law in evidence to claim compliance with a statutory requirement...'

BS 7671 is not enforceable, in that you do not have to comply with it, but in the case of non-compliance, it is the duty of the designer, installer and inspector to show how the installation complies with the legal requirements of at least three pieces of law, these being:

- The Electricity at Work regulations
- Electricity Safety, Quality and Continuity regulations
- The Building Regulations.

The Secretary of State has decreed that if it can be proven that an installation has been performed and verified in accordance with BS 7671, then the requirements of these and other pieces of law have been met. In addition to this, when an installation is inspected and tested, a certificate is issued to state that the design, installation and inspection and testing have been done in accordance with BS 7671 and that the installation complies with the requirements of BS 7671.

The understanding of the content of this standard is therefore important to everything we do within an installation.

The scope of BS 7671

Though BS 7671 applies to a large part of the industry, it does not apply to all installation work. BS 7671 only applies to electrical installations which have circuits which are supplied at nominal voltages

Character	Description
First	This number identifies the appendix in which the table or figure is located.
Second	This is a sequential letter that is used as an index to arrange tables and figures in a logical sequence.
Third	This is a sequential number that is used in conjunction with the second character as part of the sequencing.

Understanding the numbering system that is used within BS 7671 makes it easier not only to identify where each regulation, table or figure comes from within in BS 7671, but also to navigate the regulations. The regulations are arranged in alphanumerical order, so understanding the numbering system removes the need to identify page numbers. This is useful as the index does not use page numbers for identifying where items can be found, as shown in the extract below.

INDEX

It is essential that when you read regulations you also understand the context in which the regulation applies. An example of this is regulation 422.3.9 which, if not read in the correct context, would seem to imply that all systems not run in mineral insulated cable, powertrack or even busbars would have to be RCD-protected. This is not the case, as section 422.3 only relates to installations where there is a risk of fire due to the 'manufacture, processing or storage of flammable materials including the presence of dust'.

> **KEY POINT**
>
> Always check what the section is referring to when reading regulations, to ensure that you read the regulation in the correct context.

Regulation group 422.3 is typically applicable to locations such as, but not restricted to:

- barns
- woodworking facilities
- paper mills
- textile factories.

Each of these types of installation can generate large amounts of flammable dust, which can settle onto electrical machinery and, if it gets too hot, could ignite and cause a fire.

WIRING SYSTEMS AND ASSOCIATED EQUIPMENT

Chapter 52 of BS 7671 is identified as '*selection and erection of wiring systems*'. As the main function of electrical installation work is the installation of wiring systems, this is an important chapter to read and be familiar with.

BS 7671 defines a wiring system as:

> '*An assembly made up of cable or **busbars** and parts which secure and, if necessary, enclose the cable or busbars.*'

The purpose of the wiring system is to enable electrical energy to flow from the source or point of distribution to the load. In doing this, the wiring system may be subjected to factors that affect the performance and safety of the wiring system. BS 7671 defines these factors as external influences and Appendix 5 of BS 7671 classifies some of the more common factors.

These external factors, along with the safety and nature of the wiring system, need to be considered carefully, as they will all have an impact on the choices that have to be made when selecting and erecting a wiring system.

Electromagnetic effects

The first item addressed by Chapter 52 of BS 7671 is the need to be aware of and consider the electromagnetic effect of passing current through magnetic fields. The flow of electric current results in three effects:

- chemical
- thermal
- magnetic.

Assessment criteria

7.1 Specify the main requirements of the following topics in accordance with the current version of the IET wiring regulations and describe how they impact upon the installation of wiring systems, associated equipment and enclosures:

- selection and erection of wiring systems, associated equipment and enclosures.

Busbars

Busbars are normally made from copper and are solid bars, used in place of wiring, to carry larger amounts of current.

ACTIVITY

Busbars are sometimes supplied in a split-bar format. What is the advantage of this arrangement?

Faraday, Fleming and Maxwell all did studies on the flow of electric current and the resulting magnetic field. From their work, it can be shown that when current flows along a conductor, it creates a magnetic field around the conductor, that travels along the conductor in the direction of the current flow. It can also be shown that when a conductive material is placed in a moving magnetic field at right angles to the magnetic field, then an **emf** is induced into the conductive material.

emf

Electromotive force is the driving force of electrical energy and it is measured in volts.

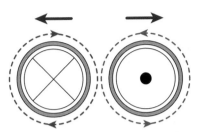

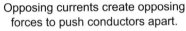

Opposing currents create opposing forces to push conductors apart.

Currents in the same direction create forces to pull conductors together.

Electromagnetic effects

If the current is flowing in the same direction in two or more conductors that are next to one another, then the magnetic fields will attracted one another and will couple together to make a stronger magnetic field, rather like putting two magnets together. The force that the magnetic field can apply to the conductor is dependent on the amount of current flowing. The more current flowing, the more magnetic force there is.

If the magnetic force is strong enough, then conductors can move when large currents flow. In the event of short circuits, earth faults and overloads, there can be large currents flowing that cause conductors to have strong magnetic forces and if the current is flowing in the same direction, conductors will slam together, crushing anything in their way.

BS 7671 Regulations 521.5.2 and 521.5.100 both address these issues:

'Single-core cables armoured with steel wire or steel tape shall not be used for an a.c. circuit.

and

Every conductor or cable shall have adequate strength and be so installed as to withstand the electromechanical forces that may be caused by any current, including fault current, it may have to carry in service.'

This includes the methods used for securing cables to cable tray or ladder racking. If a single-core armoured cable must be used in an a.c. circuit, then aluminium-armoured cable must be used.

When single-core conductors enter a **ferromagnetic** enclosure, the electromagnetic field around the conductors can cause **eddy currents** to occur within the metal of the enclosure.

Ferromagnetic

A ferromagnetic material is one that can be turned into a magnet. Typical examples include iron and steel, both of which are used to make electrical enclosures.

Eddy current

An eddy current is a circular current that is caused by the electromagnetic field of the conductor inducing an emf. These currents circulate, first one way and then the other, and can result in a lot of heat being generated.

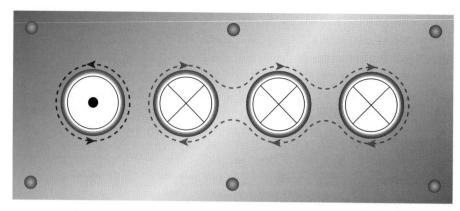

When a.c. conductors pass through a ferromagnetic plate they create
eddy currents within the metal plate which generates heat.

Eddy currents

If the eddy currents are not eliminated or reduced, the enclosure can
become hot and even electrically charged. If the wiring has been
installed, using plastic stuffing glands, then these can melt, resulting in
the enclosure cutting into the conductors and causing short circuits.

BS 7671 deals with this problem in Regulation 521.5.1 where it states:

> 'The conductors of an a.c. circuit installed in a ferromagnetic
> enclosure shall be arranged so that all line conductors and the
> neutral conductor, if any, and the appropriate protective conductor
> are contained within the same enclosure.

> Where such conductors enter a ferrous enclosure, they shall be
> arranged such that the conductors are only collectively surrounded
> by ferromagnetic material.'

One way to accomplish this is to cut a hole in the enclosure, large
enough to contain all the conductors in one opening; however, if large
single-core conductors have been used, then this may not be a
workable solution. An alternative to cutting one big hole is to cut the
holes for the glands, as normal, but then to cut a slot between
adjacent holes, joining each of the individual holes to the next.

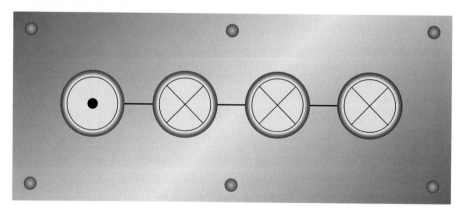

Cutting the plate between the conductors makes every conductor pass
through the same hole and therefore the eddy currents are eliminated.

Solution for eddy currents

An alternative to both of these options is to use a non-ferromagnetic plate bolted or secured to the main enclosure. The cables enter the enclosure through the non-ferromagnetic plate and therefore eddy currents are unable to form. These plates are called gland plates and are typically made from aluminium, but they can be made from rubber, plastic and even wood.

SEGREGATION

The electromagnetic effect of conductors is not just associated with electromagnetic stress. If two conductors run alongside each other and one has a large current flowing through it, this can result in an emf being induced in the second conductor. This is commonly referred to as electromagnetic interference (EMI).

EMI can cause lots of problems with modern applications and uses of electrical energy, as it can cause interference and surges or spikes within circuits. In the first amendment of BS 7671 17th edition, Section 444: Measures against electromagnetic disturbances was introduced. This new section focuses on the effect that electromagnetic disturbances can have on sensitive circuits such as those for information technology.

Some of the more common causes of these problems include:

- lightning
- switching of large loads
- short circuit faults
- rectification circuits.

These problems are, however, not only restricted to information technology systems, as electromagnetic disturbances can also have a negative effect on medical equipment.

The measures that can be adopted to ensure that the sensitive circuit is not affected include:

- screening of signal cables
- the use of surge protective devices
- installation of line, neutral and earth close together
- segregation of power circuits from more sensitive circuits
- installation of equipotential bonding networks in accordance with Regulation 444.5.3.

The segregation of sensitive circuits is covered in Regulation 444.6.1 and further guidance is given in table A444.2, which details the distance between circuits, depending on the effectiveness of the screening.

Assessment criteria

7.1 Specify the main requirements of the following topics in accordance with the current version of the IET wiring regulations and describe how they impact upon the installation of wiring systems, associated equipment and enclosures:

- segregation.

The regulation identifies that cables run through a joist of a ceiling or floor need to meet certain criteria such as:

- the cables must be run at least 50 mm from the edge of the joist that is having the nails or screws driven into it, or
- be run in either armoured or mineral insulated cables, with the armour or sheath connected to earth, or
- be run in steel conduit, which is then also connected to earth, or
- be run in steel trunking or ducting, which is then also connected to earth, or
- have additional mechanical protection on the top or bottom of each joist to prevent nails and screws from penetrating the cable, or
- be either an SELV or a PELV circuit.

The most common practice, when installing cables through joists, is to drill holes along the centre of the depth of the joist and to make sure that the top or bottom of the hole is more than 50 mm from the edge of the joist. Section 7.3.1 of the On-Site Guide (OSG) provides further guidance on the drilling of joists or the cutting of notches. The OSG takes into account the requirements of the building regulations, with regards to the size, quantity and location of holes in joists.

ACTIVITY

Access to cables installed under floorboards is often by means of a screwed trap. Why use screws rather than nails?

Notes:
1 Maximum diameter of hole should be 0.25 × joist depth.
2 Holes on centre line in a zone between 0.25 and 0.4 × span.
3 Maximum depth of notch should be 0.125 × joist depth.
4 Notches on top in a zone between 0.07 and 0.25 × span.
5 Holes in the same joist should be at least 3 diameters apart.

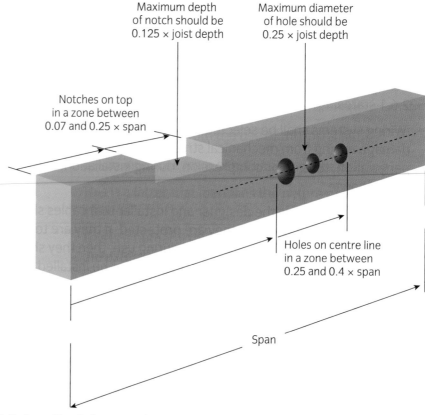

Maximum depth of notch should be 0.125 × joist depth

Maximum diameter of hole should be 0.25 × joist depth

Notches on top in a zone between 0.07 and 0.25 × span

Holes on centre line in a zone between 0.25 and 0.4 × span

Span

Details for cutting holes or notches

Size of conductors

Section 523 of BS 7671 outlines the requirements to be met when sizing cables. It covers the need to limit the conductor temperature under normal operating conditions and states the maximum operating temperatures of conductors, based on their insulating material.

Type of insulation	Temperature limit
Thermoplastic	70 °C at the conductor
Thermosetting	90 °C at the conductor
MICC that can be touched	70 °C at the sheath
MICC that cannot be touched	105 °C at the sheath

When a conductor is installed, several factors have an impact on the temperature of the conductors and the cable itself. A conductor's primary function is to carry electrical energy from the source to the destination. If the opposition to the flow of the electrical energy, the resistance, is too great, then insufficient energy will reach the load in order for it to function correctly.

Resistance is affected not only by the material, its length and its area, but also by temperature. If temperature is not considered during the design and installation stages, then the conductor may well become overloaded and overheat. Like a person doing physical exercise, a conductor has three things that can stop it cooling:

- the surrounding ambient temperature
- the closeness of other heat sources such as other loaded cables
- being wrapped in thermally insulating material.

When selecting a conductor size, these three main factors must be considered. Section 532 of BS 7671 states that Appendix 4 should be used to size the conductors. Within Appendix 4, correction factors are listed that are applied to the current rating of the circuit being designed, based on the rating of the protective device. Each factor is identified in a series of tables within BS 7671, predominately within Appendix 4. They include:

- C_a – ambient temperature
- C_c – circuits buried within the ground
- C_d – depth of burial
- C_f – use of a BS 3036 fuse
- C_g – grouping of circuits
- C_i – thermal insulation
- C_s – thermal resistivity of soil.

On pages 308 and 309 of BS 7671, the two most common formulae for sizing cables are displayed. It should be noted that Cc, Cd and Cs only apply when the cable is buried directly in the ground. The two equations are:

1   for single circuits with no grouping with other circuits

2   for circuits grouped with at least one other circuit.

Appendix 4 of BS 7671 provides a detailed approach to performing cable calculations. Further information on how to size cables, taking into account current carrying capacities and voltage drop, can be obtained from the OSG appendix F. When working out the size of a conductor, design choices are made to compensate those factors that affect the final rating and size of the cable. Any departure from this during the installation process must first be agreed with the designer of the circuit.

ISOLATION AND SWITCHING

BS 7671 identifies five different types of switching:

■ isolation

■ switching off for mechanical maintenance

■ emergency switching

■ functional switching

■ fire-fighters, switching.

Each type of switching has a different purpose and therefore a different set of requirements. These must all be understood by both the designer and the installer, so that the correct choice of switching device can be made.

Isolation

BS 7671 defines isolation as:

'A function intended to cut off for reasons of safety the supply from all, or a discrete section, of the installation by separating the installation or section from every source of electrical energy'

Under section 537.2 of BS 7671, the key requirements that need to be satisfied for isolation are identified. Regulation 537.2.1.2 states:

'Suitable means shall be provided to prevent any equipment from being inadvertently or unintentionally energised.'

This regulation is in line with the requirements of Regulation 13 of the Electricity at Work Regulations 1989, which states:

> 'Adequate precautions shall be taken to prevent electrical equipment, which has been made dead in order to prevent danger while work is carried out on or near that equipment, from becoming electrically charged during that work if danger may thereby arise.'

One simple way to achieve this is by making sure that the switch chosen for isolation can be secured in the off position by the use of a key or lock. An isolator must also break all live conductors, unless it is a TN-S or TN-C-S system and then the isolator needs only to break the line conductors. An isolator must also provide a clear indication of the position of the contacts so that it can be clearly seen when the isolator has been operated.

Isolator locked off

There are two types of isolator used within electrical installations, with the first being the more common.

- An off-load switch. This type is not able to break load current and therefore precautions should also be in place to prevent the switch from being opened during loaded conditions.
- An on-load switch. This type can make and break full-load current so, if this type of switch is used as an isolator, then no additional precautions need to be taken.

Whichever type of switch is used, it must be clearly identified by position or durable marking, to indicate the circuit that it isolates.

Mechanical maintenance

Switching off for mechanical maintenance is common on machinery that needs cleaning, lubrication or just general repair work, such as changing a lamp. The purpose of switching off for mechanical maintenance is to prevent physical injury resulting from the machine operating while it is being maintained. It is not intended to provide isolation, as there may still be live parts within the equipment.

If the switch is continuously under the control of the person performing the maintenance, then there is no requirement that the switch can be secured in the off position. However, if it is not, then such provision must be made.

A switch for mechanical maintenance must be an on-load switch, as it must be able to switch off – and keep off – the equipment it relates to. Once again, the position of the contacts needs to be clearly identified, so that it can be seen when the switch has been operated.

ACTIVITY

When installing additional remote stops in a motor control circuit, are they connected in parallel or series with the existing stop?

ASSESSMENT GUIDANCE

Remember that the input to the motor starter is still live as would be parts of the control circuit. Use the local isolator to make the starter dead.

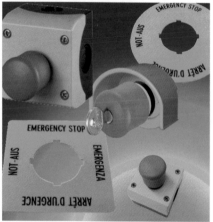

Various types of emergency stop button

A light switch

Emergency switching

The purpose of emergency switching is to enable the swift removal of any unexpected danger. Therefore, emergency switching typically takes the form of a push button that will stop the equipment it controls immediately. This does not perform the same function as isolation, as the supply will still be present to the equipment and so there will still be live conductors within the equipment.

Where possible, the operation of the emergency switch will break the supply conductors, but on larger machinery, where there will be large currents flowing in the supply conductors, this would not be practical. The emergency switching would normally be connected into the control circuit of the motor starter and operation of the emergency switching would cause the starter to become de-energised.

A device used for emergency switching should be coloured red and be set against a contrasting background; yellow is commonly used. It should also latch in the operated position and the contacts should not operate until the switch has passed the latch position. This will remove the risk of inadvertent switching if the button should be accidentally knocked. When the emergency switch is reset, the equipment it had stopped should not automatically restart.

A device for emergency switching should be located in a position where it is easy to reach and may be readily operated by anyone nearby, in the event of an emergency.

Functional switching

Functional switching is what controls the operation of the electrical equipment installed and so such switches are on-load devices. However, they do not have to be mechanical switches, as semiconductors that have no moving parts can be used for functional switching. Semiconductors cannot perform any other switching function as they do not physically break any conductors. Functional switching is concerned only with starting and stopping the current flow.

When used in circuits that perform motor control, the functional switching device or devices must be so arranged that, in the event of a loss of supply to the motor, it should stop and should not automatically restart unless a failure to restart would result in a higher level of danger.

Firefighter's switch

When electrical installations have discharge lighting circuits, or circuits with a voltage exceeding low voltage, they can present a danger to firefighters in the event of a fire. A switch is provided to enable the firefighters to switch off these circuits before they deal with the fire.

Placing out of reach

Fault protection

Fault protection is intended to provide protection for persons in the event of a fault. If a person is using the equipment as intended, then the most common fault that would cause an electric shock would be a line conductor coming in contact with the casing i.e. an earth fault. BS 7671 identifies four methods of providing fault protection:

- automatic disconnection of supply
- non-conducting location
- earth-free local equipotential bonding
- electrical separation.

The protective measure of automatic disconnection of supply is the most common protective measure used for fault protection, as it does not require any additional measures or precautions to be taken by the user. Non-conducting location, earth-free local equipotential bonding and electrical separation all require additional precautions to be made.

Automatic disconnection of supply

This is commonly referred to as ADS and BS 7671 identifies three things that need to be provided in order for ADS to be achieved. There must be a provision of:

- protective earthing
- protective equipotential bonding
- a means of disconnection.

Protective earthing is provided within an installation by the earthing conductor and the circuit protective conductors (CPC). The earthing conductor connects the distribution earth to the main earthing terminal (MET). The CPC connects the MET to the load and every accessory that forms part of the circuit.

Extraneous conductive part

A conductive part that is not part of the installation but is connected to earth and under fault conditions may be at a different potential to the rest of the metal parts within an installation.

Protective equipotential bonding is provided within an installation by the main protective bonding conductors and supplementary equipotential bonding conductors. The main protective bonding conductors connect the MET to any **extraneous conductive part**. The supplementary equipotential bonding conductors connect any simultaneously accessible parts together where they may have a difference in potential under normal or fault conditions greater than 50V a.c.

A means of disconnection is provided within an installation by a protective device such as a fuse, circuit breaker or RCBO. It does not refer to a residual current device (RCD). An RCD works on an imbalance of current between line and neutral and will not operate on overloads or short circuits. Although an RCD does provide protection for the main function of fault protection, it does not cover all aspects and so is classed as additional protection to that of basic and fault protection.

Electrical separation

One other provision that is common within some installations is the use of electrical separation for a single item of equipment, such as a shaver socket within a bathroom or washroom. The protective measure provides basic protection and fault protection by:

- basic insulation of live parts or by barriers or enclosures (Section 416 of BS 7671)
- fault protection by the provision of basic separation of the separated circuit from other circuits and from earth.

The requirements of basic protection, as outlined in Section 416, correspond to the requirements for other protective measures, such as the insulation of live parts and that all enclosures are to be to IP2X or IPXXB on the front and sides and IP4X or IPXXD for the accessible top surfaces. Fault protection is where the main differences occur, as the electrical separation from earth is achieved in most cases by the use of a transformer.

Fault protection for electrical separation

The requirements for fault protection for circuits that are using electrical separation as the protective measure are outlined within Section 413.3 and they include:

- the separated voltage is not to exceed 500 V
- live parts of the separated circuit are not to be connected to another circuit or to earth at any point
- where flexible cables are used, they must be visible if they are liable to mechanical damage
- no exposed conductive parts of the separated circuit are to be connected to the protective conductor or to earth or even to the exposed conductive parts of other circuits.

This last requirement is important to note when installing electrically separated circuits in a location containing a bath or shower, as Regulation 701.415.2 identifies the need for supplementary equipotential bonding to be installed in these locations. This must not be connected to the exposed conductive parts of an electrically separated circuit as this will render the protective measure ineffective.

How electrical separation works

As mentioned earlier, in order for a person to receive an electric shock, three criteria have to be in place, one of which was a complete circuit. In the UK all electrical installations are connected to earth at the source, so any connection to earth, within the installation, will complete a loop back to the source.

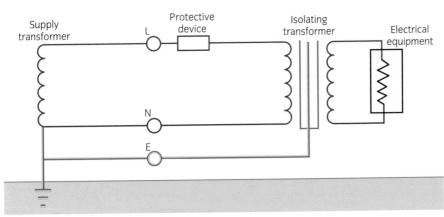

TN-S system with electrical separation

Electrical separation

In electrically separated circuits, the transformer used is an isolating transformer and this removes earth from the secondary side of the transformer. By installing the isolation transformer, there is no physical connection to a fault on the secondary side of the transformer. This results in the fault current not appearing on the primary side of the transformer and so a loop for the shock potential is not created.

Enhanced protection

There is a third form of protection that is covered by BS 7671. It focuses on the fact that some form of basic protection provides fault protection, and vice versa, without any other provisions needed. There are four forms of enhanced protection identified by BS 7671:

- SELV
- PELV
- double insulation
- reinforced insulation.

Though there are four forms, they work on two approaches, that of voltage level and that of insulation.

SELV and PELV

The use of separated extra-low voltage (SELV) and protective extra-low voltage (PELV) means that the supply cannot be above 50 V a.c. Thus, even under fault conditions, the line voltage to earth would never be above 50 V a.c. The use of these two types of system means there is never enough voltage to drive a harmful current through a person, even if they were touching the line and neutral with their bare hands.

The natural impedance of the human body means that a force greater than 50 V a.c. is required to drive a harmful current through the body.

Symbol used to represent double insulated equipment

Double or reinforced insulation

Double insulation is defined within BS 7671 as having a layer of insulation which is then covered by a second supplementary layer of insulation. In order for a person to get a shock, two layers of insulation must be damaged before live parts are exposed.

Reinforced insulation is a little bit different as there is only one layer of insulation. However, the insulation is so constructed that it requires multiple damage in order for a live part to become exposed.

Both of these forms of enhanced protection, which focus on insulation, conform to the second part of the fundamental rule of prevention of electric shock where BS EN 61140 and BS 7671 both state that protection is required under single fault conditions. If multiple faults are required to present the risk of electric shock then this has been complied with.

Assessment criteria

7.1 Specify the main requirements of the following topics in accordance with the current version of the IET wiring regulations and describe how they impact upon the installation of wiring systems, associated equipment and enclosures:

■ protection against fire.

PROTECTION AGAINST THERMAL EFFECTS (BS 7671 CHAPTER 42)

When electric current flows, heat is generated. This is a constant issue that anyone involved with electrical installation work must consider at all times. The heat generated not only impacts on how well the equipment works and the protection provided, but it can also introduce a further risk of burns or even fire. These are referred to within BS 7671 as thermal effects.

Chapter 42 of BS 7671 focuses on what steps are required to be taken to protect persons, property and livestock from:

■ the harmful effects of heat from electrical equipment
■ the ignition, combustion or degradation of materials
■ the spread of fire and smoke.

Chapter 42 also identifies the measures that need to be taken to prevent safety services from being cut off by the failure of electrical equipment.

Protection against burns

The final section within Chapter 42 addresses the considerations with regard to protecting people from getting burnt when using or touching electrical equipment. Table 42.1 identifies the maximum acceptable surface temperature of electrical equipment based on the application of the equipment and the material from which it is made.

TABLE 42.1 –
Temperature limit under normal load conditions for an accessible part of equipment within arm's reach

Accessible part	Material of accessible surfaces	Maximum temperature (°C)
A hand-held part	Metallic	55
	Non-metallic	65
A part intended to be touched but not hand-held	Metallic	70
	Non-metallic	80
A part which need not be touched for normal operation	Metallic	80
	Non-metallic	90

It is important to bear in mind that metallic items transfer heat better than non-metallic items. In addition to this, the longer that skin is kept in contact with a high temperature, the greater the damage done. In considering these two facts, the table splits the application into three groups:

■ a hand-held part, which is a part that must be touched for long periods of time to use the equipment; an example of this could be a battery-operated piece of equipment or equipment such as an electric iron

■ a part intended to be touched but not hand-held, which is a part where the time in contact with the skin is brief; an example could be a push button

■ a part not intended to be touched, which could be any part of equipment that would not normally be accessible to people.

Hand-held items that must be held to be used have the lowest maximum temperature rating. Items that are not meant to be touched are allowed the highest maximum limiting temperatures.

SPECIAL LOCATIONS (BS 7671 PART 7)

When electrical installations are designed and installed, it is important to bear in mind that, although standard external influences must be considered when selecting and erecting equipment, not all installations are the same. In fact, not all parts of an installation may be the same. BS 7671 recognises that, in some areas of an installation, there is a higher risk of electric shock due to the nature of the location or type of installation.

KEY POINT

It is important to remember that, even if a part is not intended to be touched, the higher temperature can introduce a fire risk and so the rest of Chapter 42 must also be considered.

Assessment criteria

7.1 Specify the main requirements of the following topics in accordance with the current version of the IET wiring regulations and describe how they impact upon the installation of wiring systems, associated equipment and enclosures:

■ special locations.

Special locations

BS 7671 identifies these locations and installations as special and Part 7 of BS 7671 lists them. When considering the work to be performed in a special location, or a special installation, it is important to understand the special requirements in terms of IP ratings and changes to the protection required.

Locations with baths or showers

BS 7671 does not use the term 'bathroom' as this implies only one room, and baths and showers can be found in a variety of locations, including:

- bathrooms
- shower rooms
- wet rooms
- changing rooms
- utility rooms
- bedrooms
- birthing rooms.

In fact, Section 701 applies to anywhere that a bath or shower is located, which is for general use. It does not apply to facilities that are there for emergency, such as Hazmat showers and decontamination showers.

Zones

A room containing a bath or shower has different levels of risk of electric shock in different areas of the location. BS 7671 splits the room up into different areas, based on the different level of risk. These areas are called zones and can be simplified as shown below.

Zone	Description
0	This is the area where the water is held and can be any size or shape, depending on the size and shape of the bath or basin.
1	This is the area of the perimeter of the bath or basin from the floor to a maximum height of 2.25 m. This would be the area occupied by a person if they were sitting or standing in zone 0.
2	This is the area immediately outside the perimeter of the bath or basin to a distance of 0.6 m from the bath or basin, to a maximum height of 2.25 m. This is the area that can be reached by extending an arm while sitting or standing in zone 0.

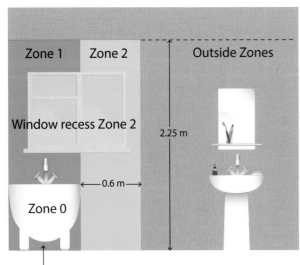

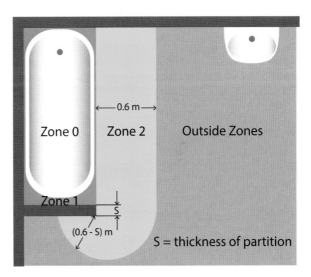

The space under the bath is:
Zone 1 if accessible without the use of a tool
outside the zones if accessible only with the use of
a tool

Extract from OSG showing zones

In some instances there may be a shower with no basin, for example, in a wet room or the showers at a sports gym. The zones in such circumstances are further clarified in BS 7671 in figures 701.1 and 701.2. If a shower head is installed at a height greater than 2.25 m from the floor, then the height of zones 1 and 2 will be increased to the height of the shower head.

Selection of equipment

Because the location has been split in zones based on risk, each zone has different requirements with regards to the equipment that can be installed within it. The main focus in this location is the presence of water and so the IP ratings of equipment and accessories are important.

Zone	Equipment requirements
0	All equipment must have a minimum IP rating of IPX7.
	No switches or controls are to be installed in this zone.
	Only permanently connected, purpose built equipment that operates on SELV with a voltage not exceeding 12 V a.c. rms may be installed.
1	All equipment must have a minimum IP rating of IPX4.
	Only switches and controls of SELV circuits with a maximum voltage of 12V a.c. rms are allowed.
	Only permanently connected, purpose-built equipment such as electric showers, heated towel rails, luminaires, may be installed unless they are protected by SELV or PELV and have a voltage not exceeding 25V a.c. rms.
2	All equipment must have a minimum IP rating of IPX4, except shaver-supply units complying with BS EN 61558-2-5, as long as they are installed where they are not likely to get wet from showers.
	Only switches and controls of SELV circuits are allowed.
	Only permanently connected, purpose-built equipment may be installed.

BS 1361 13A socket outlets are not permitted within a horizontal distance of 3 m from the edge of zone 1.

Other key points to consider

Other considerations that need to be made when performing electrical installations within a location containing a bath or shower include the need for all circuits within the location to be RCD-protected, in accordance with Regulation 701.411.3.3. This requires an RCD with a maximum rating of 30 mA and a maximum disconnection time of 40 ms when an imbalance of five times the rating of the RCD is flowing between line and neutral.

There is also a requirement for supplementary equipotential bonding to be installed between all exposed metal work of the accessories

Supplementary equipotential bonding

fitted and extraneous conductive parts. This is to maintain the potential between these parts to 50 V a.c. or less. However, Regulation 701.415.2 makes it clear that this is not always required if the location is within an installation which already has main protective bonding installed. In order to omit supplementary equipotential bonding three criteria must be met and these are:

- all final circuits within the location must meet the maximum disconnection times stated within Table 41.1 by having an earth fault loop impedance, Zs, within the limits specified in Tables 41.2 to 41.4 of BS 7671, and

- all final circuits of the installation must have additional protection in the form of an RCD, and

- all extraneous-conductive-parts must be effectively connected to the main protective bonding.

The first two requirements are quite simple to establish but the third option requires a confirmation by testing. Using continuity test method 2 (from Guidance Note 3), the resistance between the metal part of the location and the MET can be measured. If the reading is low enough, then supplementary equipotential bonding is not required. The note below Regulation 701.415.2 identifies how to establish this.

On a 30 mA RCD, the maximum resistance before supplementary bonding is required is 1667 Ω. If a reading above this value is obtained then supplementary equipotential bonding is required. If a reading above 1 MΩ is obtained then it can be confirmed that supplementary equipotential bonding is not required as the metal part of the location is insulated from earth and therefore is not an extraneous-conductive-part.

Other special locations

As mentioned earlier, BS 7671 covers many other special locations and installations including:

- swimming pools and basins (702)
- sauna heaters (703)
- construction and demolition sites (704)
- agricultural and horticultural premises (705)
- conducting locations with restricted movement (706)
- caravan and camping parks (708)
- marinas (709)
- medical locations (710)
- exhibitions, shows and stands (711)
- solar PV (712)

ACTIVITY

Name one type of physical problem with electrical ceiling heating.

ASSESSMENT GUIDANCE

There are different types of RCD. Make sure you know the difference between the types.

Assessment criteria

7.1 Specify the main requirements of the following topics in accordance with the current version of the IET wiring regulations and describe how they impact upon the installation of wiring systems, associated equipment and enclosures:

- flammable/explosive atmospheres.

- mobile and transportable units (717)
- caravans and motor homes (721)
- gangways (729)
- fairgrounds, amusement parks and circuses (740)
- electric floor and ceiling heating (753).

Each special installation or location requires careful consideration with regards to how the installation is designed and installed. It is important for both the designer and installer to be familiar with the additional special requirements as specified in the relevant section of BS 7671.

Note: Amendment 3 of BS 7671 also includes outdoor and low-voltage lighting as special installations.

Common key points to consider include the requirements for a higher level of IP rating than normal, the provision of additional protection by supplementary equipotential bonding and RCDs and the use of SELV or PELV with restricted values.

FLAMMABLE AND EXPLOSIVE ATMOSPHERES

Locations where there is a higher risk of fire or explosion are not listed as special installations or locations within Part 7 of BS 7671. However, Regulation 110.1.3 states that reference should be made to BS EN 60079 when installing electrical equipment in explosive gas atmospheres.

The types of location commonly identified with flammable or explosive atmospheres include fuel-filling stations. Appendix 2, Section 5, of BS 7671, provides information on further legislation that should be read and applied when working in these types of installation.

'5. For installations in potentially explosive atmospheres reference should be made to:

i   the Electricity at Work Regulations 1989 (SI 1989 No 635)

ii   the Dangerous Substances and Explosive Atmospheres Regulations (DSEAR) 2002 (SI 2002 No 2776)

iii the Petroleum (Consolidation) Act 1928

iv   the Equipment and Protective Systems Intended for Use in Potentially Explosive Atmospheres Regulations 1996 (SI 1996 No 192)

v   relevant British or Harmonised Standards'

Section 5, of BS 7671, goes on to say that, under the Petroleum (Condensation) Act 1928, local authorities have the authority to grant licences for premises where fuel is stored. This means that the local authorities also have the power to add conditions to the licence that affect what can be installed, and how. Before undertaking any electrical installation work in a filling station, it is essential to be familiar with the requirements as set by the local authorities. Guidance on how to comply can be obtained from the Energy Institute publication, Design, Construction, Modification, Maintenance and Decommissioning of Filling Stations.

Filling station forecourt

It would be easy to think that for flammable and explosive atmospheres, the requirements of Regulation group 422.3 would apply; however, Regulation 422.3 states:

> '...This regulation does not apply to selection and erection of installations in locations with explosion risks, see BS EN 60079-14 and BS EN 61241-14.'

It is therefore essential to refer to BS EN 60079 for the selection and erection of electrical equipment, especially with regards to the installation of switchgear within these locations. This is because, when switchgear is operated, arcs are formed by the switching and in a flammable or explosive atmosphere this could cause ignition and fire, if not an explosion. Flameproof fittings must be used throughout and, when inspecting and testing, reference must also be made to BS EN 61241-17 to ensure safety at all times.

Remember: work in these locations requires experience and knowledge of the dangers that exist. This type of work would normally be undertaken by specialists.

ASSESSMENT CHECKLIST

WHAT YOU NOW KNOW/CAN DO

Learning outcome	Assessment criteria	Page number
1 Understand the procedures, practices and statutory and non-statutory regulatory requirements for preparing work sites for the installation of wiring systems and associated equipment	*The learner can:*	
	1 Explain the health and safety requirements and legal duties of employers and employees in establishing a safe working environment	2
	2 Interpret relevant sources of information which will inform installation work	6
	3 Specify the actions required to ensure that electrical installation work sites are correctly prepared in terms of health and safety considerations	11
2 Understand the procedures for checking the work location prior to the commencement of work activities	*The learner can:*	
	1 State the preparations that should be completed before electrical installation work starts	15
	2 Explain how to check for any pre-existing damage to customer/client property and state why it is important to do this prior to commencement of any work activity	21
	3 State the actions that should be taken if pre-existing damage to customer/client property is identified	21
	4 Specify methods for protecting the fabric and structure of the property before and during installation work	27

Learning outcome	Assessment criteria	Page number
3 Understand the practices, procedures and regulatory requirements for completing the safe isolation of electrical circuits and complete electrical installations	*The learner can:*	
	1 Specify and undertake the correct procedure for completing safe isolation with regard to:	28
	■ carrying out safe working practices	
	■ correct identification of circuit(s) to be isolated	
	■ identifying suitable points of isolation	
	■ selecting correct test and proving instruments in accordance with relevant industry guidance and standards	
	■ correct testing methods	
	■ selecting locking devices for securing isolation	
	■ correct warning notices	
	■ correct sequence for the safe-isolation of an electrical circuit and complete electrical installation	
	2 State the implications of carrying out safe isolations to:	28
	■ other personnel	
	■ customers/clients	
	■ public	
	■ building systems (loss of supply)	
	3 State the implications of not carrying out safe isolation to:	28
	■ self	
	■ other personnel	
	■ customers/clients	
	■ public	
	■ building systems (presence of supply)	

Learning outcome	Assessment criteria	Page number
4 Understand the types, applications and limitations of wiring systems and associated equipment	*The learner can:*	
	1 State the constructional features, applications, advantages and limitations of types of cable	40
	2 State the constructional features, applications, advantages and limitations of types of cable and conductor containment systems	40
	3 Describe how environmental factors can affect the selection of wiring systems, associated equipment and enclosures	40
	4 State the types of wiring systems and associated equipment used for: ■ lighting systems ■ power systems (final circuits) ■ distribution systems (sub mains) ■ environmental control/building management systems ■ emergency management systems ■ security systems – fire alarm/prevention; unlawful entry; emergency lighting ■ closed circuit TV, communication and data transmission systems	40

Learning outcome	Assessment criteria	Page number
5 Understand the procedures for selecting and using, tools, equipment and fixings for the installation of wiring systems, associated equipment and enclosures	*The learner can:*	
	1 State the procedures for selecting and safely using appropriate hand tools, power tools and adhesives for electrical installation work	91, 101, 108
	2 State the procedures for selecting and safely using equipment for measuring and marking out for wiring systems, equipment and enclosures	89, 101
	3 State the criteria for selecting and safely using tools and equipment for fixing and installing wiring systems, associated equipment and enclosures	91, 101, 108
	4 State the criteria for selecting and safely using fixing devices for wiring systems, associated equipment and enclosures, giving consideration to	101, 110
	■ load bearing capacity	
	■ fabric of structure	
	■ environmental considerations	
	■ aesthetic considerations	
6 Understand the practices and procedures for installing wiring systems, associated equipment and enclosures	*The learner can:*	
	1 Specify and apply the installation methods and procedures to ensure that in accordance with the installation specification and statutory and non-statutory regulations:	115
	■ wiring systems, enclosures, cables and components are securely fixed and installed	
	■ a wiring system's mechanical integrity is maintained	
	■ no damage to the wiring system or its components has occurred	
	2 Specify methods and techniques for restoring the building fabric	115

Learning outcome	Assessment criteria	Page number
7 Know the regulatory requirements which apply to the installation of wiring systems, associated equipment and enclosures	*The learner can:* **1** Specify the main requirements of the following topics in accordance with the current version of the IET wiring regulations and describe how they impact upon the installation of wiring systems, associated equipment and enclosures: ■ selection and erection of wiring systems, associated equipment and enclosures ■ isolation and switching ■ protection against fire ■ protection against electric shock ■ special locations ■ segregation ■ flammable/explosive atmospheres	122

Multiple-choice questions

From part 1

1 Which one of the following is covered by BS 7671, IET Wiring Regulations?

   a) systems for the distribution of energy to the public

   b) equipment of aircraft

   c) public premises

   d) electric fences

From part 2

2 *A final circuit arranged in the form of a ring and connected to a single source of power* is the definition of:

   a) ring final main circuit

   b) ring main circuit

   c) radial ring spur circuit

   d) ring final circuit.

From part 3

3 A three-phase four-wire TT system achieves an earth connection via:

   a) a water pipe

   b) a gas pipe

   c) the consumer's earth electrode

   d) the supply neutral.

From part 4

4 The protective measure *placing out of reach* defines *arm's reach* as:

   a) 1.5 m

   b) 2.5 m

   c) 3.5 m

   d) 4.5 m.

From part 4

5 In a correctly designed circuit, the relationship between the protective device rating and the design current will be:

   a) $Ib \leq In$

   b) $Ib \leq Iz$

   c) $It = I\,Ib$

   d) $In \leq Ib$?

From part 4

6 Discrimination between protective devices of the same type, is normally achieved by a multiplier of:

   a) 1.5

   b) 1.75

   c) 2

   d) 3?

From part 5

7 Which one of the following may be used as an isolator?

   a) 2 gang lighting switch

   b) switch disconnector

   c) pull-cord switch

   d) push button

From part 5

8 Which one of the following is not permitted as a protective conductor?

   a) steel armouring of SWA cable

   b) copper water utility pipe

   c) MICC copper sheathing

   d) galvanised conduit

From part 5

9 A fireman's switch shall be installed at a
   maximum height of:

   a) 1.4 m

   b) 2.5 m

   c) 2.75 m

   d) 3.75 m?

From part 6

10 The minimum insulation resistance and test
    voltage for a 400 V circuit is:

    a) 0.25 MΩ and 400 V

    b) 0.5 MΩ and 500 V

    c) 1MΩ and 500 V

    d) 1 M and 1000 V?

From part 7

11 In a sauna with an electric heater, the heater is
    placed in:

    a) H zone

    b) Zone 1

    c) Zone 2

    d) Zone 3

From appendices

12 The addition of a socket to an existing circuit
    would require the issue of a:

    a) Schedule of Inspections

    b) Schedule of Test Results

    c) Electrical Installation certificate

    d) Minor Electrical Installation works
        certificate?

UNIT 306
Understanding the principles, practices and legislation for the termination and connection of conductors, cables and cords in electrical systems

This unit is designed to enable learners to understand and interpret the principles, practices and legislation associated with the termination and connection of conductors, cables and cords in electrotechnical systems. Its content is the knowledge needed by a learner to underpin the application of skills for terminating and connecting conductors, cables and cords in electrotechnical systems, in accordance with statutory and non-statutory regulations/requirements.

LEARNING OUTCOMES

There are three learning outcomes to this unit:

The learner will:

1 understand the principles, regulatory requirements and procedures for completing the safe isolation of electrical circuits and complete electrical installations

2 understand the regulatory requirements and procedures for terminating and connecting conductors, cables and flexible cords in electrical wiring systems and equipment

3 understand the procedures and applications of different methods of terminating and connecting conductors, cables and flexible cords, in electrical wiring systems and equipment.

This unit will be assessed by means of an assignment consisting of:

- Practical cable installation exercise covering assessment criteria AC 3.5

- Oral questions during practical exercise covering assessment criteria AC 2.1; 2.2; 2.3; 3.1; 3.2; 3.3; 3.4.

- Carry out safe isolation (Common Task) covering assessment criteria AC 1.1; 1.2; 1.3

Understand the principles, regulatory requirements and procedures for completing the safe isolation of electrical circuits and complete electrical installations

Assessment criteria

1.1 State the implications of carrying out safe isolations

1.2 State the implications of not carrying out safe isolations

1.3 Specify and undertake the correct procedure for completing safe isolation

KEY POINT

Just because it looks 'OFF' don't assume it is. Always test to make sure.

SmartScreen Unit 306
Handout 2

ISOLATION PROCEDURE

Safe isolation includes using notices like these

A procedure is a series of steps taken to accomplish an outcome. Safe isolation procedure requires that a series of actions is carried out in a set order.

Reasons for isolation

Isolation is defined in BS 7671 as 'a function intended to cut off for reasons of safety the supply from all, or a discrete section, of the installation by separating the installation or section from every source of electrical energy'. The primary reason for isolation is to remove the risk of electric shock, fire and burns when working on the electrical system.

The impact of failure to carry out safe isolation

People often think, incorrectly, that failure to isolate an installation only affects the person working on the electrical installation. However, the table shows that the effects of failure to isolate can be far reaching.

Areas/people affected	Hazards to consider
Operative working on the system	Risk of electric shock when working on system
Other personnel such as co-workers	Risk of electric shock from contact with live parts when barriers are removed
Customers and clients	Risk of electric shock Damage to equipment if short circuits are created when working on live equipment
General public	Risk of electric shock from contact with live parts when barriers are removed or due to unfamiliarity with building
Building services	Short circuits created when working on live equipment may damage building systems equipment, resulting in damage to the building fabric due to heat, for example

The impact of carrying out isolation

Although isolation removes the risk of electric shock, the act of isolation can have a negative impact on other users of the installation and, in some cases, can give rise to additional hazards. Careful planning of isolation can minimise the risk of harm from these hazards. The table indicates some of the possible hazards that should be considered when planning an isolation.

ACTIVITY

You cannot tell if an electrical circuit is dead or alive just by looking at it. What type of test instruments should be used to test the supply?

Areas/people affected	Hazards to consider
Other personnel such as co-workers	Unexpected loss of power to machines, giving rise to dangerous situations Loss of lighting that may be required to carry out normal operations
Customers and clients	Loss of service will affect normal operations, for instance the loss of supply to tills and card machines will stop shop sales being made
General public	Loss of lighting, emergency lighting and fire alarms (both fire alarms and emergency lighting systems have battery back-up systems but these run down over time)
Building services	Unexpected loss of power to computers, resulting in data loss Loss of communications systems Loss of services such as heating and ventilation Some processes may require shut-down and restart procedures

Before carrying out isolation, it is important that any hazards that may arise due to the isolation are identified. A risk assessment should be carried out and a method statement should be agreed. It is imperative that permission is gained from the person responsible for the building's operation *prior* to carrying out isolation.

Members of the public do not always react as expected, so you must secure the work area and post appropriate signs and warning notices.

Equipment required for secure isolation

The following equipment is required to enable isolation to be carried out:

- *voltage indicator*, complying with the requirements of GS38
- *proving unit*, or known source to prove the voltage indicator works
- suitable *lock* with single key, to be used to secure the means of isolation and to guard against inadvertent re-energising of the supply
- *notices* to inform and warn
- hand tools such as *screwdrivers* are needed to gain access to equipment.

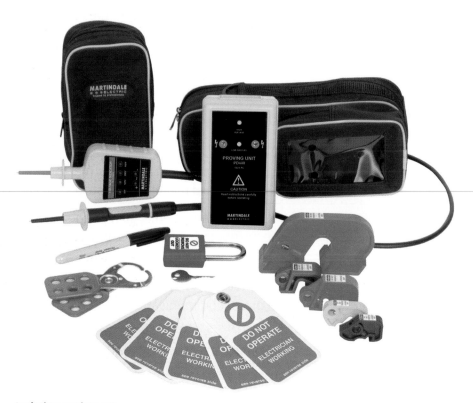

Isolation equipment

The isolation procedure

The order of actions to be taken in the isolation procedure is as follows:

1 Select an approved voltage indicator to GS38 and confirm operation.

2 Locate correct source of supply to the section needing isolation.

3 Confirm that the device used for isolation is suitable and may be secured effectively.

4 Power down circuit loads if the isolator is not suitable for on-load switching.

5 Disconnect using the located isolator (from step 2).

6 Secure in the off position, keeping the key* on person, and post warning signs.

7 Using the voltage indicator, confirm isolation by checking ALL combinations.

8 Prove the voltage indicator is still operational by testing on a known source such as proving unit.

* If the device is a fuse or removable handle (instead of a lockable device), keep this securely under supervision while work is undertaken.

1 Select an approved voltage indicator to GS38 and confirm operation

GS38, 'Electrical test equipment for use by electricians' is published by the Health and Safety Executive (HSE). An electronic version can be downloaded free of charge from the HSE website.

This is a non-statutory document which offers advice on the selection and use of the following items for circuits with rated voltages not exceeding 650 V:

- test probes
- leads
- lamps
- voltage-indicating equipment
- measuring equipment.

When selecting and using test equipment, it is important that you select items which comply with the requirements of GS38. Bear in mind that not all test equipment meets these requirements, for instance multi-meters may not come supplied with leads that comply with GS38; it would be the responsibility of the person ordering the equipment to specify these.

The requirements of GS38 are that:

- probes
  - have finger barriers to stop inadvertent touching of live parts
  - are insulated to leave an exposed tip of no more than 4 mm
  - be fitted with high breaking capacity (HBC) fuses or be fitted with current limiting impedances
- leads
  - are adequately insulated
  - are coloured to distinguish one lead from another
  - are flexible
  - are sheathed to provide mechanical protection
  - are of the correct length for the purpose
  - do not have accessible exposed parts
- terminations and sockets
  - are shrouded.

Voltage indicators, used for *detecting* voltage as opposed to voltmeters that are used for *measuring* voltage, fall into two categories:

- *Test lamps* – which rely solely on an illuminated bulb. These are required to have overcurrent protection, usually provided by means of a 500 mA HBC fuse. The test leads and probes must meet the requirements discussed above and the leads must be held captive by the instrument, rather than being detachable.

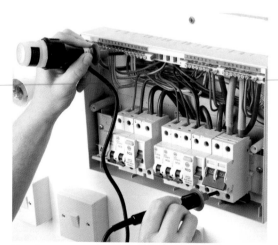

Test lamps

- *Detectors* – which use two or more separate indicating devices, usually visual and audible devices. Detectors of this type that have current limiting built into them are not required to have overcurrent fuses fitted but must meet all the other requirements for leads and probes. Once again the leads cannot be detachable.

It is very important that all combinations of conductors are checked.

1 For single phase, three individual tests are used:

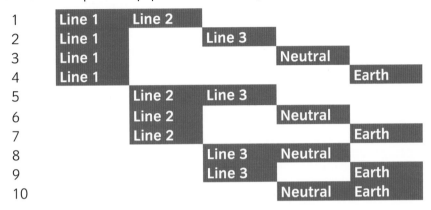

1	Line	Neutral	
2	Line		Earth
3		Neutral	Earth

2 For three-phase equipment or circuits, there are 10 individual tests:

1	Line 1	Line 2		
2	Line 1		Line 3	
3	Line 1			Neutral
4	Line 1				Earth
5		Line 2	Line 3	
6		Line 2		Neutral
7		Line 2			Earth
8			Line 3	Neutral
9			Line 3		Earth
10				Neutral	Earth

Never omit any of the tests; even the neutral–earth test is important. It is not uncommon to come across installations where the neutral is from one circuit and the line from another. This incorrect practice is known as 'borrowed' or 'shared neutrals' and is highly dangerous; isolating the line conductor could leave the neutral at a dangerous potential. The neutral–earth test identifies such circuits.

Place warning notices

Warning notices advise other people about what has been done. A notice on an isolator informs others that the operative has switched off because they are working on the installation and that the isolation has taken place for safety reasons.

Warning notice

> **KEY POINT**
>
> Always assume that equipment and circuits are live until confirmed dead by means of tests.

> **ACTIVITY**
>
> Some equipment will still be live even after it has been switched off. Can you think of two such pieces of equipment?

SmartScreen Unit 306
Worksheet 2

Isolation procedure

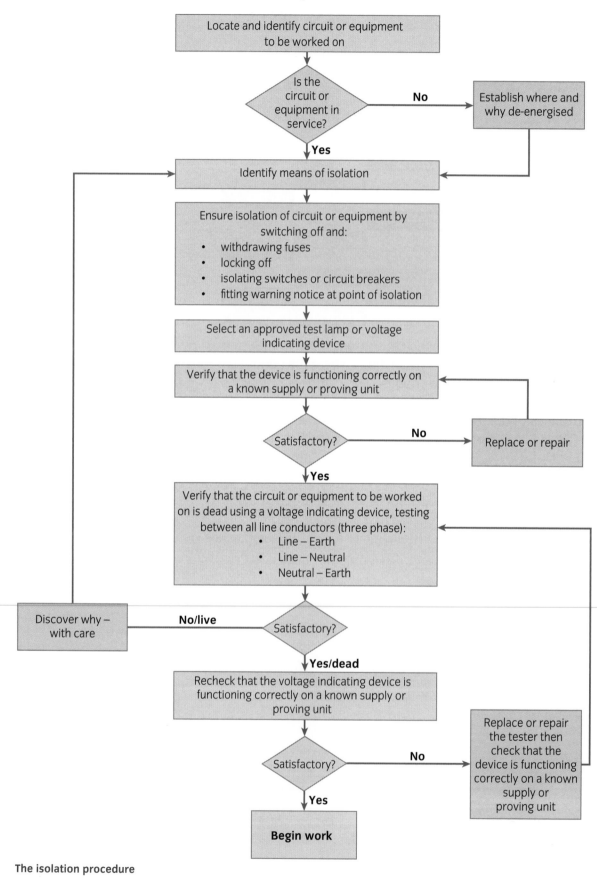

The isolation procedure

Understand the regulatory requirements and procedures for terminating and connecting conductors, cables and flexible cords in electrical wiring systems and equipment

SOURCES OF INFORMATION RELEVANT TO TERMINATING AND CONNECTING CONDUCTORS

When carrying out the termination and connection of conductors, cables and flexible cords in electrical wiring systems and equipment, a number of sources of information are available to the installer. Sources of information include:

- statutory documents
- codes of practice
- British standards
- IET wiring regulations
- manufacturers' instructions
- installation specifications.

Assessment criteria

2.1 Identify and interpret appropriate sources of relevant information for the termination and connection of conductors, cables and flexible cords in electrical wiring systems and equipment

SmartScreen Unit 306
PowerPoint presentation 5

Sources of Information

Statutory documents

The two key statutory documents when terminating and connecting cables and equipment are:

- The Health and Safety at Work (etc) Act 1974 (HASAWA)
- The Electricity at Work Regulations 1989 (EAWR).

The Health and Safety at Work (etc) Act 1974 (HASAWA)

The HASAWA is the primary health and safety document and as such applies to all work activities. The HASAWA is considered to be an 'umbrella Act' or an 'enabling Act'. Many other health and safety documents sit underneath this Act. Whilst they are not Acts of Parliament, they are 'enabled' to be statutory by this Act. These 'other' regulations are specific and cover all work activities including, asbestos, noise, electricity, personal protective equipment etc.

The Electricity at Work Regulations 1989 (EAWR)

The Electricity at Work Regulations is one of the statutory documents that sit under the 'umbrella' of the HASAWA.

Regulation 3 – Deals with duties placed upon individuals, including:

- employers
- employees
- the self-employed.

The levels of 'duty' under law depend on the work being carried out, but will apply when carrying out any electrical work.

Regulation 4 – Calls for 'danger' to be avoided at all times. This means that correct methods and procedures must be followed at all times, so that danger cannot arise.

Regulation 5 – *'No electrical equipment shall be put into use where its strength and capability may be exceeded in such a way as may give rise to danger'*

By definition, 'electrical equipment' in the EAWR includes the conductors. 'Strength and capability' needs to take into account not only the load current carrying capabilities of a conductor, but also the conductor's ability to handle fault currents. Terminations and joints must not introduce 'weaknesses' in the conductor path.

Regulation 6 – *'Electrical equipment which may reasonably foreseeably be exposed to:*

a) *mechanical damage;*

b) *the effects of the weather, natural hazards, temperature or pressure;*

c) *the effects of wet, dirty, dusty or corrosive conditions; or*

d) *any flammable or explosive substance, including dusts, vapours or gases,*

shall be of such construction or as necessary protected as to prevent, so far as is reasonably practicable, danger arising from such exposure.'

The application of suitable glands for the environment are applicable to this regulation. Environmental factors and the application of the IP code will be discussed later.

Equipment must be suitable for the environment

Regulation 7 – 'All conductors in a system which may give rise to danger shall either…

a) be suitably covered with insulating material and as necessary protected so as to prevent, so far as is reasonably practicable, danger; or

b) have such precautions taken in respect of them (including, where appropriate, their being suitably placed) as will prevent, so far as is reasonably practicable, danger'.

Whilst this regulation deals with insulation, any termination method that could result in damage to insulation in the future would be in direct contravention of this regulation. It is important that cables are terminated in such a way as to ensure that insulation is not damaged, or will not become damaged in the future, due to abrasion or overheating.

Regulation 8 – Relates specifically to earthing, but contains the requirement that the conductors be of *'sufficient strength and current-carrying capability to discharge electrical energy to earth.'*

Once again, consideration needs to be given to strength and capability. In this case it is the earthing conductors, including joints, that need to be considered, to ensure that they are adequate to deal with fault currents and to make sure that poor termination does not introduce high resistances into the earth fault path.

Regulation 9 – *'If a circuit conductor is connected to earth or to any other reference point, nothing which might reasonably be expected to give rise to danger by breaking the electrical continuity or introducing high impedance shall be placed in that conductor unless suitable precautions are taken to prevent that danger.'*

ACTIVITY

With reference to Regulation 9 of the EAWR, what would be the danger associated with a disconnected neutral in the distribution board?

Supply isolated – ready for work to start

This regulation applies to the neutral conductor of a system. It is obvious that the placing of a single pole switch in the neutral of a circuit is unsafe, but this regulation would also be applicable when considering the connection of neutrals, so as to ensure that they cannot become inadvertently disconnected.

Regulation 10 – *'Where necessary to prevent danger, every joint and connection in a system shall be mechanically and electrically suitable for use.'*

This regulation applies specifically to terminations and connections and applies to both live and protective conductors. Connections in live conductors must be able to carry the steady load current without overheating either the insulation of the cable or the terminal it is connected to. Both live conductors and protective conductors must be able to carry fault currents safely. An unsound connection poses a real hazard to persons and property.

Regulation 11 – Deals with overcurrent protection.

Regulation 12 – Deals with the provision of suitable isolation devices.

Regulation 13 – Deals with taking precautions to avoid danger and highlights the need for isolation before carrying out any electrical installation work.

Regulation 14 – is concerned with 'working on or near live conductors'. This regulation gives three criteria that must be met before 'working on or near live conductors' is defendable under law. The three criteria are:

a) 'it is unreasonable in all the circumstances for it to be dead

b) it is reasonable in all the circumstances for a person to be at work on or near a conductor while it is live

c) suitable precautions (including where necessary the provision of suitable protective equipment) are taken to prevent injury.'

ASSESSMENT GUIDANCE

It is not clever to work on live equipment: it is downright dangerous. Just because someone finds it inconvenient to turn the supply off is no reason to leave it on.

Suitable precautions being taken

Whilst there are occasions, such as testing, when working near live conductors is required, there is **never** a time during installation or terminating conductors when the above criteria can be met, and therefore live working is not allowed.

Regulation 15 – Calls for adequate working space, access and lighting. This regulation should be considered in all work activities relating to electricity and especially so when terminating cables and conductors.

Regulation 16 – States 'No person shall be engaged in any work activity where technical knowledge or experience is necessary to prevent danger or, where appropriate, injury, unless he possesses such knowledge or experience, or is under such degree of supervision as may be appropriate having regard to the nature of the work.'

This regulation calls for anyone carrying out electrical work to be competent. This is also called for in BS 7671 (skilled persons). If a person does not have the necessary technical knowledge or experience for a particular task, they should not be undertaking that task unless they are supervised by someone who has the necessary knowledge or experience.

The above is a discussion of the statutory requirements for working on or alongside electrical systems and has a direct bearing on the termination of cables and conductors.

ACTIVITY

What is the significance of Regulation 29 of EAWR?

Codes of practice

Approved codes of practice (ACoP)

Approved codes of practice (ACoP) provide practical guidance on how to comply with Health and Safety regulations. ACoPs are not law, but do have a special legal status. Following the guidance contained within an ACoP, means that duty holders can be confident that the law has been complied with.

Other codes of practice

The IET publish codes of practice relevant to electrical work and include such titles as:

- Code of Practice for In-Service Inspection and Testing of Electrical Equipment
- Code of Practice for Electric Vehicle Charging Equipment Installation.

Codes of practice may include information relevant to termination and connection of conductors and cables. The Code of Practice for In-Service Inspection and Testing of Electrical Equipment, for example, contains information on the connection of flexes to plugs.

ASSESSMENT GUIDANCE

Testing is one area where live working may be required. It is difficult to carry out an Earth loop impedance test without the supply.

British Standards

A number of British standards relevant to electrical equipment exist and these are listed in Appendix 1 of BS 7671. These may be a source of information for the termination of cables and conductors.

IET Wiring Regulations – BS 7671 IET Requirements for Electrical Installations (Formerly the IEE Wiring Regulations)

These regulations are non-statutory, but it is accepted by the Health and Safety Executive (HSE), that compliance with the IET Wiring Regulations should give compliance with the requirements of the Electricity at Work Regulations 1989.

BS 7671 is the national standard for electrical installation work and includes regulations specific to the termination of cables and conductors. These are mainly contained in Section 526, which is discussed later in the book.

Alongside BS 7671 is a set of guidance notes that provide detailed interpretation and guidance on compliance with the various parts of the Wiring Regulations. The specific Guidance Notes are:

1 Selection and Erection
2 Isolation and Switching
3 Inspection and Testing
4 Protection Against Fire
5 Protection Against Electric Shock
6 Protection Against Overcurrent
7 Special Locations
8 Earthing and Bonding.

IET Guidance Notes

Variation order

A variation order, also referred to as an architect's instruction (AI), is issued by the person ordering the work, whether that be an architect, a consultant, a main contractor or even the client, if they are project-managing the job for themselves.

A variation order is an official acknowledgement of changes and/or additional costs that can be made to the original contract.

THE EFFECT OF THE ENVIRONMENT ON THE EFFECTIVENESS OF TERMINATION

When terminating cables, the environmental conditions need to be taken into account, to ensure that the terminations are sound, both mechanically and electrically.

Damp or wet conditions

Termination of cables in damp or wet conditions may result in premature failure of the connection. The insulation used in mineral insulated cables is **hygroscopic**, meaning it absorbs moisture. Therefore, if the cable is terminated in a damp environment there is a possibility that moisture may be sealed in the cable, resulting in a fault. The armouring of cables also needs to be considered, as any moisture that is trapped in the termination, possibly by the shroud, could lead to corrosion of the armour.

Steps need to be taken to ensure that damp is kept out of terminations

Assessment criteria

2.3 Describe methods and procedures appropriate to the installation environment to ensure the safe and effective termination and connection of conductors, cables and flexible cords in electrical wiring systems and equipment

ACTIVITY

How could the risk of water contamination to cable termination be minimised?

ASSESSMENT GUIDANCE

Gaskets are often fitted to prevent the entry of moisture. Make sure they are fitted correctly.

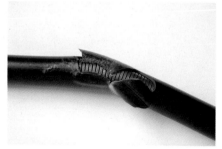

Low temperature may affect items such as PVC conduit

Temperature

Temperature also needs to be considered when terminating cables. At temperatures below 0°C, PVC becomes brittle and may split or crack when bent of flexed. High temperatures may also pose a problem, as the PVC sheath of the cables may become more elastic and stretch. This could become a problem when the conductors cool, as shrinkage may occur, leaving an excess amount of conductor exposed, and thus leading to possible faults, or even the risk of electric shock due to exposed live parts.

Dust and foreign bodies

Fibre-optic cables need to be terminated in a dust-free environment. Specks of dust trapped in the termination could result in data loss during use. Dust and powders could become a fire or explosion risk and so, in environments where there is a high dust content, care needs to be taken to ensure that these cannot enter electrical equipment.

Special care needs to be taken in environments containing gases or vapours that are flammable.

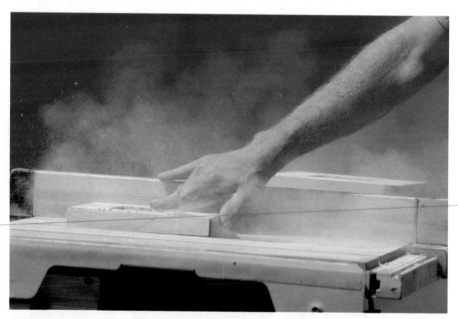

Dust could be a problem when terminating fibre optic cables

Understand the procedures and applications of different methods of terminating and connecting conductors, cables and flexible cords in electrical wiring systems and equipment

TERMINATING CABLES AND BS 7671

Section 526 of BS 7671 is the key section relating to the terminating and connection of conductors. Below is a regulation-by-regulation discussion of the requirements.

526.1

This requires that all connections shall provide 'durable electrical continuity and adequate mechanical strength and protection'. This is a general requirement and is in keeping with requirements of the EAWR. The subsequent regulations give further information on how this can be achieved.

526.2

This requires that *'the means of connection shall take account of, as appropriate:*

 i *the material of the conductor and its insulation*

 ii *the number and shape of the wires forming the conductor*

 iii *the cross-sectional area of the conductor*

 iv *the number of conductors to be connected together*

 v *the temperature attained at the terminals in normal service, such that the effectiveness of the insulation of the conductors connected to them is not impaired*

 vi *the provision of adequate locking arrangements in situations subject to vibration or thermal cycling.*

Where a soldered connection is used, the design shall take account of **creep**, *mechanical stress and temperature rise under fault conditions.'*

There are a large number of considerations covered by this regulation. Take each of these points in turn.

The material of the conductor and its insulation

Different metals or, more correctly, dissimilar metals may react with each other, resulting in corrosion of the termination. It is therefore important to make sure that compatible materials are used when terminating cables and conductors. A further discussion of corrosion takes place later in the book.

SmartScreen Unit 306
Handout 10

Solder creep

Sometimes called 'cold flow', 'solder creep' can occur when the termination is under constant mechanical stress and the solder can literally move or 'creep'. Incidence of creep increases with temperature.

Tube lug and bell mouth lug designed to capture all strands of conductor

Terminals that are not large enough to accommodate all the strands are problematic

Ceiling rose showing linked terminals

The number and shape of the wires forming the conductor

Conductors come in many formats, round or triangular, solid or stranded. It is important to select terminations that are compatible with the cable and/or conductor. Failure to use compatible parts may result in the conductor becoming loose within the termination.

When crimping lugs onto cables, it is important that the correct size lugs and crimp dies are used, to ensure that a sound mechanical and electrical connection is made. Shaped lugs are available to use with triangular and half-round conductors used in some two-, three- and four-core cables. Where flex is being terminated, it is important that the ends are treated, for example, fitting ferrules to ensure that the individual strands are not spread and are all contained within the termination.

The cross-sectional area of the conductor

The cross-sectional area (csa) of the cable was carefully selected at the design stage, to ensure that the current-carrying capacity of the conductor was adequate for the circuit load. When terminating the cable, it is important to ensure that all strands of the conductor are contained within the terminal, to ensure that the current-carrying capacity is maintained. Terminals need to be large enough to house the conductor and to be suitably rated to carry the circuit load. Failure to meet either of these requirements could result in overheating at the termination that, in turn, may cause damage to the equipment and/or the insulation of the cable. This in turn may pose a fire risk.

The number of conductors to be connected together

The termination must be suitable for the number of conductors to be connected together. Attempting to fit more conductors into a terminal than it is designed to hold will invariable result in one or more of those conductors not being properly connected and becoming loose over time. In the case of accessories, where more than one conductor is intended to be connected, manufacturers will provide a number of linked terminals to house all of the conductors so that a sound mechanical and electrical connection is made.

The temperature attained at the terminals in normal service, such that the effectiveness of the insulation of the conductors connected to them is not impaired

The maximum operating temperature of conductors with thermoplastic insulation (usually PVC) is 70 °C, whilst the maximum operating temperature of thermosetting cables (XLPE) is 90 °C. When using thermosetting cables at 90 °C, it is important to make sure that the terminals are capable of withstanding this temperature, as the majority of electrical accessories are designed to operate at 70 °C. Other specialist cables may operate at higher temperatures, so it is important to check the maximum operating temperature of the terminals.

The provision of adequate locking arrangements in situations subject to vibration or thermal cycling

Where conductors are terminated into machinery, this can cause vibration and may adversely affect the terminations. In such cases, it is common practice to use, for the final connection, one of these options:

- a flexible cable – where added mechanical protection is required, braided flex such as SY flex can replace standard flex

- a flexible conduit – where unsheathed cables make the final connection or where additional mechanical protection is required, the cables can be housed within flexible conduit

- an anti-vibration loop in the cable – with cables such as MICC, or other cable types which are not flexible, a loop is included in the cable to allow the cable to absorb any vibration.

ASSESSMENT GUIDANCE

The use of a flexible connection will also allow the motor to be moved when adjustment or alignment is required.

Motor connected by means of SY flex

Motor connected by means of flexible conduit

Motor connection showing anti-vibration loop

Thermal cycling is heating and cooling of in this context, metal, which causes expansion and contraction and, eventually, the loosening of terminals. It is greatest when cables are run at or near their maximum operating temperature. The effects can be reduced by ensuring that terminations and connections are kept tight.

526.3

Poor and loose terminations cause many fires of electrical origin. Regulation 526.3 requires every connection or joint to be accessible for inspection, testing and maintenance, with the exception of:

- joints designed to buried in the ground

- a compound-filled joint

- an encapsulated joint

- a cold tail of a floor or ceiling heating system

- a joint made by welding or soldering

- a joint made with an appropriate compression tool

- spring-loaded terminals complying with BS 5733 and marked with the symbol (MF).

ACTIVITY

How can a lighting circuit be designed so that any connections are always accessible?

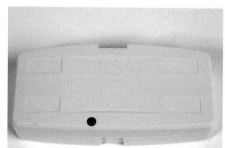

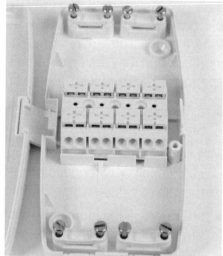

Types of connection not required to be accessible

526.4

This requires that the insulation of the cable must not be adversely effected by the temperature attained at a connection. An example of this may be where a cable with an insulation temperature rating of 70°C is connected to a bus-bar which has been designed to run at a higher temperature. In this case the insulation on the cable would be removed from the cable to a suitable distance and replaced with insulation capable of withstanding the higher temperature.

526.5

All terminations and joints in **live conductors** must be enclosed within a suitable enclosure or accessory. There are no exceptions. This requirement applies to both low-voltage and extra-low voltage connections, but sadly it is not uncommon to see poor examples of connections, especially where down-lighters are fitted.

Example of an acceptable connection

In harmony with this regulation is regulation 421.1.6, which requires that all enclosures have suitable mechanical and fire-resistant properties.

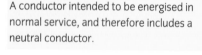

ASSESSMENT GUIDANCE

From this regulation it is clear that the practice of using screw terminal junction boxes and placing these in inaccessible places, such as ceiling or floor voids, is not in compliance with BS 7671.

Live conductor

A conductor intended to be energised in normal service, and therefore includes a neutral conductor.

526.6

This requires that there is *'no appreciable mechanical strain in the connections of conductors.'* Mechanical strain may come about due to:

- the conductor bending too tightly before entering the terminal, causing the termination to be under constant stress. Cables must be installed in accordance with the minimum bend radii, as given in the IET On-Site-Guide or IET Guidance Note 1 (GN1).

- cables having no form of strain relief fitted. This can exert mechanical forces on the termination, due to the weight of the cable pulling on the termination. Cables should be fixed at the maximum distances given in the IET On-Site-Guide or IET GN1. The fitting of suitable cable glands, where cables enter enclosures, provides strain relief on the terminations.

- cables terminated without any slack. Under faulty conditions large electromagnetic forces, due to high fault currents, are exerted on the cables, with the highest forces being exerted at the 'crutch point', where cables come together at the outer sheath. If there is little slack in the cables, the forces will be transferred to the terminal.

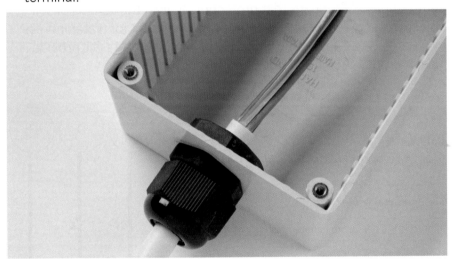

Using a gland to provide strain relief

526.7

This requires that, where a joint in a conductor is made within an enclosure, the enclosure must provide adequate mechanical protection as well as protection against relevant external influences. The minimum requirements for an enclosure to meet the requirements for basic protection are that:

- the bottom sides and face meet at least IP2X or IPXXB

- for the top surface, the enclosure must meet IP4X or IPXXD.

However it may also be necessary to take into account external influences such as water ingress or dust ingress, depending on the location.

ACTIVITY

Is it permissible to use junction boxes in Zone 0 and Zone 1 of a swimming pool?

The IP Code

The IP Code is an international code specifically aimed at manufacturers of enclosures and equipment. It applies to degrees of protection provided by electrical equipment enclosures with rated voltages not exceeding 72.5 kV. The abbreviation 'IP' stands for **International Protection**, so the full title is International Protection Code (IP). The code is defined in IEC 60529 (BS EN 60529).

The three general categories of protection given in the standard are:

1 the ingress of solid foreign objects (first digit)
2 the ingress of water (second digit)
3 the access of persons to harmful electrical or mechanical parts.

When referring to the IP code in wiring regulations, 'X' is used in place of the first or second numeral, to indicate that:

1 the test is not applicable to that enclosure **or**
2 in the case of standards, the classification of protection is not applicable to this standard.

For example, IP2X means that protection against the ingress of solid objects must meet at least IP2 but the requirement for water ingress protection is not applicable in this case. Manufacturers will provide a full code, such as IP44, for the enclosure.

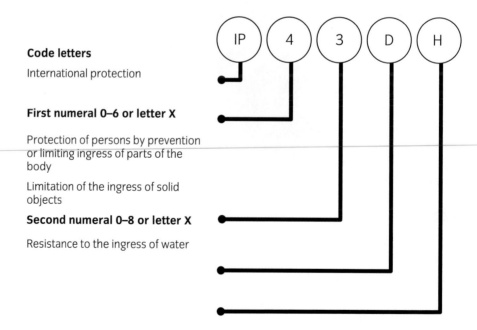

Code letters

International protection

First numeral 0–6 or letter X

Protection of persons by prevention or limiting ingress of parts of the body

Limitation of the ingress of solid objects

Second numeral 0–8 or letter X

Resistance to the ingress of water

How the IP Coding system works

First numeral 0–6: Ingress of solid objects

IP	Requirement	Example
0	No protection	
1	Full penetration of **50.0 mm** diameter sphere not allowed and shall have adequate clearance from hazardous parts. Contact with hazardous parts not permitted.	Back of the hand
2	Full penetration of **12.5 mm** diameter sphere not allowed. The jointed test finger shall have adequate clearance from hazardous parts.	A finger
3	The access probe of **2.5 mm** diameter shall not penetrate.	A tool such as a screwdriver
4	The access probe of **1.0 mm** diameter shall not penetrate.	A wire
5	Limited ingress of dust permitted. No harmful deposit.	
6	Totally protected against ingress of dust.	Dust-tight

The first digit relates to the ingress of solids

ASSESSMENT GUIDANCE

Do not use your own finger to test equipment. Use a proper test finger.

Second numeral 0–8: Ingress of water

IP	Requirement	Example
0	No protection	
1	Protected against vertically falling drops of water	
2	Protected against vertically falling drops of water with enclosure tilted 15° from the vertical	
3	Protected against sprays to 60° from the vertical	
4	Protected against water splashed from all directions	Outdoor electrical equipment
5	Protected against low-pressure jets of water from all directions	Where hoses are used for cleaning purposes
6	Protected against strong jets of water	Where waves are likely to be present
7	Protected against the effects of immersion between 15.0 cm and 1.0 m	Inside a bath tub
8	Protected against submersion or longer periods of immersion under pressure	Inside a swimming pool

The second digit relates to the ingress of water

ACTIVITY

What is meant by the code IP44?

Additional letter A–D: Enhanced protection of persons

IP	Requirement	Example
A	Penetration of 50.0 mm diameter sphere up to guard face must not contact hazardous parts	The back of the hand
B	Test finger penetration to a maximum of 80.0 mm must not contact hazardous parts	A finger
C	Wire of 2.5 mm diameter × 100.0 mm long must not contact hazardous parts when spherical stop face is partially entered	A screwdriver
D	Wire of 1.0 mm diameter × 100.0 mm long must not contact hazardous parts when spherical stop face is partially entered	A wire

The IP codes that you are most likely to come across are:

1 For protection against the ingress of solid objects and protection to persons. The codes are IP2X and IP4X used in relation to barriers and enclosures.

2 For protection against the ingress of water the codes are IPX4, IPX5, IPX6, IPX7 and IPX8.

3 For enhanced personal protection IPXXB and IPXXD again used in relation to barriers and enclosures.

Cable entry at electrical accessory not meeting the IP code.

Non-sheathed cables outside of an enclosure

526.8

Where the sheath of a cable has been removed the cores of the cable must be enclosed within an enclosure as detailed in 526.5. This also applies to non-sheathed cables, contained within trunking or conduit.

526.9

The group of regulations designated 526.9 relates to the connection of multi-wire, fine-wire and very-fine wire conductors.

526.9.1

To stop the ends of multi-wire, fine-wire and very-fine wire conductors from spreading or separating, this regulation requires that suitable terminals, such as plate terminals, or suitable treating of the ends be undertaken. One suitable method is to fit ferrules on the ends of the conductor. Manufactures will almost always fit some form of ferrule to the ends of flexes so that the conductor can be terminated in a screw terminal.

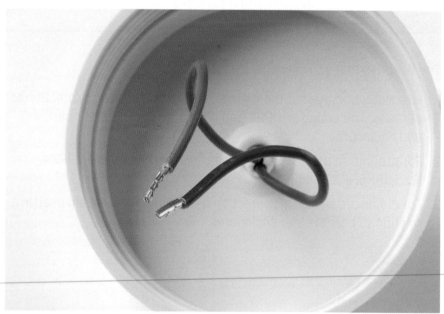

Ferrule fitted to pendent flex

ASSESSMENT GUIDANCE

Soldering flex forms a hard mass that, when subjected to vibration, may work loose. Most appliances come fitted with a 13A plug but occasionally, a flex with soldered ends is supplied. In this situation, advice from the manufacturer should be followed.

ACTIVITY

Why should solid 2.5 mm² not be twisted together when terminated at 13A sockets?

526.9.2

Soldering or tinning of the ends of multi-wire, fine-wire and very-fine wire conductors is not permitted if screw terminals are used.

526.9.3

The connection of soldered and non-soldered ends on multi-wire, fine-wire and very-fine wire conductors is not permitted where there is relative movement between the two conductors.

CONNECTION METHODS

Assessment criteria

3.1 Explain the advantages, limitations and applications of various connection methods

Allowable connection methods include:

- screw
- crimped
- soldered
- non-screw compression.

Each method has its advantages and disadvantages.

Screw terminals

When cables are terminated into electrical equipment, the type of terminal used must be taken into account. Most accessories use a grub screw, which is screwed down onto the conductor to ensure it is retained in place. Problems arise, however, when the terminal is designed to take more than one conductor, or a conductor of a larger CSA than the one being installed.

The common types of screw terminal used in the accessories within electrical installations are:

- square
- circular
- moving-plate
- insulation-displacement
- pillar.

Square base terminal

The use of square terminals can be seen in accessories such as socket outlet face-plates, where two or even three cables can be terminated in the same terminal. These terminals are designed to accept up to three cables and, where a single conductor is used, the screw can miss or damage the conductor. To minimise the potential for problems, the end of the conductor is bent over to increase the contact area available for the screw of the terminal.

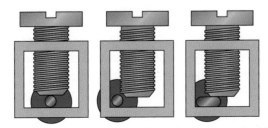

Square terminal showing possible problems if conductor is not doubled over

Bending over the end of a conductor: bend is too big, correct and too small

Circular base terminal

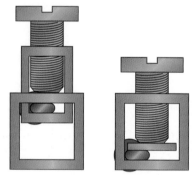

Two types of moving-plate terminal

To bend over the end of the conductor, it is necessary to remove twice the normal amount of insulation from the cable. Then a pair of long-nosed pliers is used to fold the exposed conductor in half. It is important to make sure that both sides of the bend are the same length, as shown in the centre diagram below. If the return edge is too long, as shown in the left diagram below, the conductor may protrude from the terminal, causing a shock hazard. If it is too short, as shown on the right diagram below, the bend will be redundant.

Circular terminal

Circular terminals can be seen in accessories such as light switches and ceiling roses, where single cables or small cables are terminated. They are also commonly found in consumer units and distribution boards on both the neutral and earth bars.

Circular bottom terminals are designed to ensure that the conductor is positioned directly beneath the terminal screw, so there is no need to bend the end of the conductor over.

Moving-plate terminal

Moving plate terminals are often used on protective devices, such as circuit breakers and fuse holders that are mounted in consumer units and distribution boards.

The option of bending over the end of the conductor depends not only on the size of the conductor, but also on the size and type of the terminal. If the terminal is the type where the bottom moves up towards the top when the screw is tightened, it is not necessary to bend over the conductor, as the terminal tightens evenly. If, however, the terminal has a plate that moves towards the bottom of the terminal as the screw is tightened, small conductors should be doubled over. If there is any doubt, refer to the manufacturer's recommendations.

No matter what type of terminal is being used, different conductor types should never be mixed within the same terminal. If flexible cable is terminated within the same terminal as solid or stranded cable, the screw may fail to clamp on the flexible cable and only a few of the fine strands may be secured. This could result in a poor electrical connection and the wire might come loose.

If there is no option other than to mix flexible cable and solid or stranded conductors in the same terminal, the flexible conductor must be fitted with a ferrule. This is a small, metal tube that is crimped onto the end of a flexible conductor to hold the strands together.

Terminating copper and aluminium conductors within the same terminal should also be avoided due to the electrolytic reaction between the two different metals.

Whatever types of terminal and conductor are being used, always make sure that the screw tightens on the conductor and not the insulation. To ensure this, the insulation should stop at the opening of the terminal. Take care not stop the insulation too early, leaving the conductor exposed, with the possibility of faults occurring.

Advantages and disadvantages of screw terminals

The advantages of screw terminals are:

- they are cheap to produce
- they are reliable
- they are easily terminated, with basic tools
- the terminals are reusable.

The disadvantage of screw terminals are that:

- over-tightening could result in damage to the terminal or the conductor
- under-tightening of the terminals can result in overheating and arcing
- terminals can become loose, due to movement of the conductor in use or due to mechanical vibration
- terminations need to be accessible for inspection.

ASSESSMENT GUIDANCE

Common faults with terminations are exposed conductors or screwing down onto the insulation.

Crimps

Crimps and crimp lugs come in two basic forms, insulated and uninsulated.

Insulated crimps

When using a cable crimp lug, the wire's insulation must be stripped back about 5 mm. This enables the crimp to be installed with the correct amount of conductor within the crimp and with the insulating section being sealed down onto the insulation of the wire.

Crimp lugs come in three colours for the different size of wires:

- red – 1 mm^2 to 1.5 mm^2 wires
- blue – 1.5 mm^2 to 2.5 mm^2 wires
- yellow – 4 mm^2 to 6 mm^2 wires.

Insulated crimps

The jaws of the crimping tool are shaped to apply a different crimp style and pressure to the conductor and insulation sides of the connection. The crimping tool applies the correct amount of pressure through a ratchet that cannot be defeated unless the correct amount of pressure is used, or the release button is pressed.

Once a crimp has been installed, it must be checked to ensure that the conductor of the wire protrudes from the crimped part of the lug and that the insulation has been trapped on the other end, so that no exposed conductor is showing. Bearing in mind that crimps are used in applications where vibration occurs, they should never be used on solid conductors.

Uninsulated crimps

Uninsulated crimps are used on conductors with cross-sectional areas from 6 mm^2 upwards. It is important that the crimp is sized in accordance with the cross-sectional area of the conductors and is compatible with the conductor material being terminated.

Whilst a hand crimper may be suitable for smaller-sized conductors, on larger conductors a hydraulic crimper will be required. Battery-operated crimp tools are available that take all of the hard work out of crimping a lug onto a conductor.

Method

Select the correct size cable lug for the conductor. Cable lugs with different-sized holes are available, so ensure always chose one that has the correct-sized hole for the connection bolt or screw.

ACTIVITY

It is not unknown for pliers to be used for fitting crimps, instead of the correct crimping tool. List two possible faults that could occur.

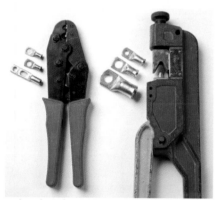

Uninsulated crimps

Make sure the appropriate lug is selected

Strip enough insulation from the cable so that the copper conductor meets the end of the cable lug, while the sheath of the cable fits tight to the base of the lug.

Measure how much to strip

Ensure that the cable reaches right to the end of the lug tube.

Fitting the lug to the conductor

The cable lug is then crimped to the cable, using a proprietary cable crimper that suits the size of the lug.

Lug being crimped to the conductor

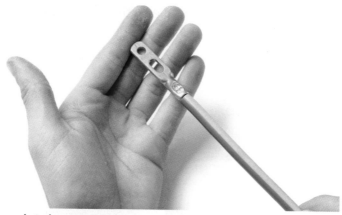

Conductor ready to be connected

Advantages and disadvantages of crimped connections

The advantages of crimped connections are that they:

- are quick and convenient to install
- provide a secure termination
- do not need to be accessible for inspection.

The disadvantages of crimped connections are that:

- special tools are required
- tools for larger sizes are expensive to purchase
- crimps cannot be reused.

Soldered terminations

In the past, lugs were soldered to cables, but this has mainly been replaced by the use of crimp lugs.

Soldering is mainly used in the assembly of electronic components and equipment.

Soldering is mainly used with electronic equipment

Advantages and disadvantages of soldered terminations

The advantages of soldered terminations are that:

- They provide a good electrical connection
- They offer good mechanical strength
- large numbers of connections can be made within a small area as the joint area is very small.

The disadvantages of soldered terminations are that:

- a heat source is required
- there are hazards associated with molten metals
- there may be damage to conductor insulation
- there may be damage to components.

Non-screw compression

Non-screw compression connectors, including push-fit connectors, have been used in many associated industries such as lighting manufacturing for a number of years and have proved to be robust and reliable, both electrically and mechanically. In recent years these have been used more and more in electrical installations and, depending on choice of connector, can be used for joining:

- solid conductors to solid conductors
- flexible conductors to solid conductors
- flexible conductors to flexible conductors.

SmartScreen Unit 306
Handout 12

A wide range of push fit connectors is available, to cope with various cable types

Manufacturers make a range of accessories to go alongside these connectors that ensure the termination of cables can be both speedy and reliable.

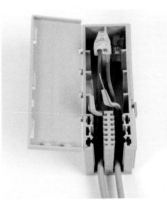

Termination of PVC cables into purpose-designed box

Advantages and disadvantages of non-screw compression terminations

The advantages of non-screw compression terminations are that:

- they are quick and convenient to install
- they provide a secure termination
- they are not affected by vibration
- no special tools are required
- they are reusable
- they are maintenance free.

The disadvantage of non-screw terminations are that:

- they are generally not available for cable sizes exceeding 6 mm^2.

PROVING THAT TERMINATIONS AND CONNECTIONS ARE ELECTRICALLY AND MECHANICALLY SOUND

It is important that when terminations are complete, they are verified to be both electrically and mechanically sound. The procedures to follow will include both inspection and testing.

Inspection will include checks such as:

- making sure that all terminations are tight. This includes both live and protective conductors and can be accomplished by both careful scrutiny and by 'tugging' the conductor to ensure it is securely fastened in the termination
- checking visually that the electrical connection is made to the conductor rather than the insulation
- checking visually that conductive parts are not accessible to touch
- checking that the correct termination methods have been used and that they are suitable for:
  - □ the type of conductors being terminated
  - □ the environment in which the termination is to be used.

KEY POINT

Careful and thorough inspection will identify the majority of faults with terminations.

Testing will need to be carried out to ensure that:

- conductors are continuous
- there are no shorts in the conductors
- conductors are connected to the correct points.

The appropriate tests would be:

- continuity, including that of the cpc and ring final circuit conductors
- insulation resistance
- polarity and phase rotation.

It is important that correct inspection and testing procedures are followed.

THE CONSEQUENCES OF TERMINATIONS NOT BEING ELECTRICALLY AND MECHANICALLY SOUND

If the termination of cables and conductors is not electrically or mechanically sound, the consequences can be disastrous. The cause of terminations not being electrically and mechanically sound is usually high-resistance joints or corrosion.

The effects of high-resistance joints

The most common cause of high-resistance joints is a loose connection. When current is passed across such a joint, it heats up. This is likely to cause damage to the cable insulation and/or the connected electrical equipment. In the worst case, this may result in the overheating of adjacent material, resulting in a possible fire. It is important to make sure that cables are seated properly in the terminals and that the terminals are correctly tightened. Manufacturer's instructions need to be consulted to check whether a torque setting is given for connections, which must then be complied with.

It should be remembered that, even with a sound connection, when current is flowing, the conductors and terminations will heat up, resulting in expansion of the metal, which can lead to loosening of the terminal. This is why terminals, apart from those exempted by regulation 526.3, must be accessible for maintenance and inspection.

Another cause of loose terminations is vibration from such things as machinery. It is important that initial terminations are correctly made and tightened and that regular maintenance is carried out to ensure that loose connections cannot occur.

Corroded terminals will also result in high-resistance joints.

The effects of corrosion

Corrosion is typically defined as the breaking down or destruction of a material, especially a metal, through chemical reactions. The most common form of corrosion is rusting, which occurs when iron combines with oxygen and water. Most metals, with the exception of precious metals such as gold and platinum, do not exist in metallic form in nature, but rather exist as ore.

For corrosion to occur four conditions must exist.

1 There must be an anode (corroding) and a cathode (protected) component.

2 There must be an electrical potential between the anode and the cathode.

Assessment criteria

3.3 Explain the consequences of terminations not being electrically and mechanically sound

ACTIVITY

What type of tape should be used to repair damaged insulation?

Lug bolted to casing of equipment

ASSESSMENT GUIDANCE

Don't forget other types of connection that are likely to be subjected to adverse conditions. These would include earth electrode connections.

3 The anode and cathode must be connected by a metallic path of low resistance.

4 The anode and cathode must be immersed in an electrolyte, which is an electrically conductive fluid.

When two dissimilar metals, such as aluminium and brass, are in contact with one another points 1, 2 and 3 are met. In the presence of a fluid such as moisture, corrosion will occur. The tendency to corrode can be reduced by plating terminals or using a corrosion inhibitor on the jointed parts. Bimetallic crimps are available for joining aluminium to copper. These are engineered to obtain the best possible transition between the metals and corrosion is reduced by coating the inside of the crimp with grease.

HEALTH AND SAFETY REQUIREMENTS APPROPRIATE TO TERMINATING AND CONNECTING CONDUCTORS, CABLES AND FLEXIBLE CORDS IN ELECTRICAL WIRING SYSTEMS AND EQUIPMENT

When terminating cables and conductors, as with all electrical work, it is essential to:

- ensure adequate lighting, access and working space
- follow correct isolation procedures
- secure the work area with barriers
- avoid creating hazards, such as trip and slip hazards
- work in accordance with risk assessments and method statements
- use appropriate PPE such as:
  □ hard hats
  □ boots
  □ hi-vis vests
  □ gloves
  □ knee pads
  □ safety glasses
  □ correct clothing for the environment
- ensure that appropriate tools are used for the job and that tools are maintained in good order
- make sure you know what the job involves
- make sure you have the knowledge and experience for the task being undertaken.

Individual requirements specific to the cables being terminated will be listed under each cable type.

TECHNIQUES AND METHODS FOR THE SAFE AND EFFECTIVE TERMINATION AND CONNECTION OF CABLES

The following section will describe and illustrate the preparation of cables for connection. Various methods will be illustrated. The health and safety requirements for each method and the tools required to terminate will be discussed for each type of cable. The cables covered in this section are:

- thermosetting insulated cables, including flexes
- single and multicore thermoplastic (PVC) and thermosetting insulated cables
- PVC/PVC flat profile cable
- MICC (with and without PVC sheath)
- SWA cables (PILC, XLPE, PVC)
- armoured/braided flexible cables and cords
- data cables
- fibre optic cable
- fire-resistant cable.

Terminating some of these cables requires the use of glands and shrouds as described below.

Cable glands

Cable glands are available in a range of sizes and formats and with a bewildering array of designatory letters and numbers: BW, CW, CX, CXT to name but a few. So what do all these letters mean? These tables provide the answers.

Table 1 – First letter

Code	Definition
A1	For unarmoured cable with an elastomeric or plastic outer sheath, with sealing function between the cable sheath and the sealing ring of the cable gland.
A2	As type A1, but with seal protection degree IP66 – means 30 bar pressure
B	No seal
C	Single outer seal
E	Double (inner & outer) seal

Assessment criteria

3.5 Interpret and apply the techniques and methods for the safe and effective termination and connection of cables

SmartScreen Unit 306
Handout 11

ASSESSMENT GUIDANCE

PILC (paper insulated lead covered) is limited to older distribution systems and should rarely be seen by a contracting electrician, except where it comes through a floor to a service head. PVC, XLPE and other types of armoured cables are now used in place of PILC.

SmartScreen Unit 306
Handout 1

Table 2 – Second letter

Code	Designation of cable armouring
W	Single wire armour
Y	Strip armour used
X	Braid
T	Pliable wire armour

From the tables, these meanings can be gathered.

BW   A gland without seals suitable for single wire armour cable. SWA for indoor use.

CW   A gland with a single outer seal for single wire armour cable. SWA for outdoor use.

CX   A gland with a single outer seal for braided cables such as SY flex for outdoor use.

Additional letters may be used to signify a method of termination, such as in the designation 'CXT'. In this case the third letter, T, signifies that the braid is formed into a tail (T) and the gland is designed to terminate by this means.

The construction and fitting of each type of gland will be discussed separately, alongside each cable termination type.

From left to right BW, CW, CX and CXT gland

When cables are terminated for use in potentially explosive atmospheres, glands must be suitable for the environment and comply with ATEX or IECEx standards. Operatives must be trained and competent to **CompEx Scheme** standards.

CompEx Scheme

The CompEx Scheme is now the global solution for validating core competency of employees and contract staff of major users in the gas, oil and chemical sectors. This covers both offshore and onshore activities.

What is the purpose of a cable gland?

A gland can be used to:

- maintain the IP rating of an enclosure
- provide continuity of earth
- provide strain relief to terminations.

A gland is an integral part of the termination of a cable and so must be fitted correctly. Incorrect fitting could result in:

- water being allowed to enter an enclosure
- the connection to earth not being adequate and posing a shock risk in the event of an earth fault
- strain being placed on cables and the cables pulling out of terminals, creating either a short-circuit fault or an earth fault.

ACTIVITY

Identify the tool shown.

Shrouds

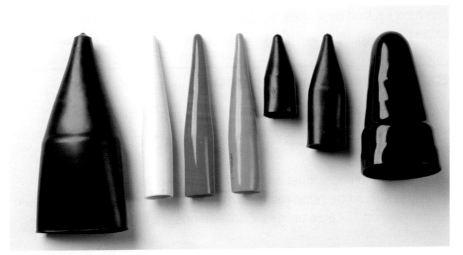

A range of shrouds

What does a cable shroud do? A cable shroud can aid the process of keeping the surface of the gland clean and free from the build-up of dirt. It does not, however, necessarily improve the ingress protection (IP) rating of the cable gland. In fact, the gland will invariably have been tested and rated without the installation of a cable shroud. The shroud may provide corrosion protection to cable armour or the sheath, but it must be installed in such a way as not to trap moisture under itelf and thus increase the corrosion potential. If the shroud is too loose on the cable sheath, moisture may enter the assembly and, as the fit with the gland is going to be tight, the moisture will be trapped.

Shrouds generally come in PVC or LSF (low smoke and fume) varieties, in a range of sizes to match the gland, and in a range of colours to match the sheath colour of the cable, with black being the most common.

A badly fitted shroud could trap moisture

How to fit a shroud to ensure a tight fit to the cable sheath

The same method of fitting a shroud can be used with all cables.

STEP 1 Push the shroud lightly on to the cable so that a small bulge appears where the cable end is. Do not push too hard, as this will stretch the shroud and you will end up cutting in the wrong place.

STEP 2 Cut the shroud at the bulge, with a pair of side cutters or, better still, a pair of cable croppers.

STEP 3 Push the shroud onto the cable. The top of the shroud should now be a snug fit to the outer sheath of the cable. Remember, when assembling the gland and shroud combination, the shroud goes on before the gland.

The next section describes the common methods of terminating cables.

Cable entry to an enclosure

When a cable enters an enclosure, the integrity of the enclosure should not be compromised. The entry may have to meet one or more of these criteria.

■ The point of entry must not cause damage to the cable. Rough edges should be removed, as a minimum, rubber grommets should be used on all cable entries. Cable glands are a better alternative.

Rubber grommet to enclosure. Grommets protect cables from rough edges

■ The entry of the cable should not compromise the IP rating of the enclosure. For basic protection this is:

- □ top surface IP4X – a 1mm diameter wire will not enter
- □ front, sides and bottom, a 12.5 mm diameter object will not enter.

Non-compliance on cable entry

- ■ There may be IP ratings that are applicable for the ingress of water:
  - □ for an enclosure outside a building, it is likely to be IPX4 splash proof
  - □ for an enclosure where water jets are used, the rating should be IPX5.

There may be requirements for fire protection. Where there is a fire risk due to powders or dust being present in locations such as a carpenter's workshop, the minimum IP rating is IP5X.

Splash proof socket

Terminating flexes

Tools required

- ringing tool (Method 1)
- stripping knife (Method 2).

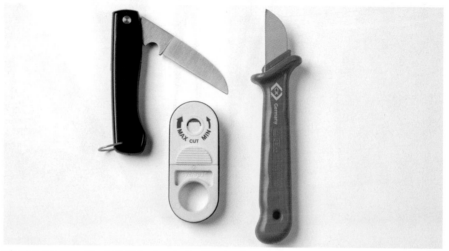

Ringing tool (Method 1) Stripping knife (Method 2)

Safety considerations

- Cuts to hands from use of knife

The use of gloves and eye protection is recommended.

Method

Before the flex can be terminated, the outer sheath must be removed.

Removal of the outer sheath

Two methods of removing the outer sheath are outlined here.

Method 1

The outer sheath can be removed with the use of a ringing tool. These come in various shapes and forms, the most basic of which is shown in the diagram. This tool slides over the end of the cable to the required stripping position, and is then rotated around the cable, cutting it slightly.

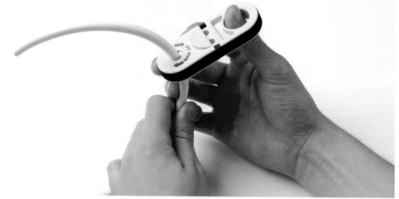

Flexible cable sheath stripping tool for use on flex

The ringing tool is removed, the cable is bent to finish the cut and the sheath is then pulled off.

Outer sheath being removed

Method 2

There may be times when a ringing tool is not available, so it is important to know how to remove the outer sheath without this tool. Thermoplastic has a tendency to split when it is damaged and pressure is applied, and this can be used to help strip the sheath.

Bend the cable into a tight bend at the point where the sheath is to be removed. Using a sharp knife, gently score the top of the bend. The thermoplastic will split open like a little mouth.

Gently work the split until the inner cables are visible and then unbend, rotate and rebend the cable at about 90° from the first point. Repeat the scoring of the sheath until again the wires become visible. Repeat the bend and cut until the sheath can be removed, being careful, at all times, not to cut into the wires.

Bending cable to cause split in sheath

Once the sheath has been removed, the wires can be stripped and connected, as detailed in the next section, but remember that the conductors are fine strands and therefore need to be fully contained with a suitable terminal or they need to treated in some way, such as fitting ferrules, to make them stable.

Fire-resistant cable

Tools required

- stripping knife
- spanners or grips to suit gland size.

Safety considerations

- Cuts to hands from use of knife.

The use of gloves and eye protection is recommended.

ACTIVITY

Use sources of information, such as wholesale catalogues, to find out about different types of fire-resistant cable.

Completed FP termination

Fire-resistant cables are stripped in a similar way to flexes. However, take care with the insulation on the inner cores as this is usually silicon rubber rather than PVC, and is easily damaged. Entry to enclosures is usually by means of a plastic gland.

Terminating single-core cables

Tools required

- Cable strippers.

Different types of cable strippers are available

Safety considerations

There is a risk of injury from slipping with cutters or pliers. The use of gloves and eye protection is recommended.

Method

As single-core cables do not have outer sheaths to remove because they are intended to be installed within trunking or conduit, the termination method is straightforward and requires the minimum of tools. This termination method will therefore also apply to the final connection of other cable types.

Cable stripper jaws

Use of wire-strippers is recommended for single-core conductors. Automatic wire-strippers tend to rip the insulation from the conductor and can damage it. Manual wire-strippers are preferred and these come in various forms, but all work on a similar principle.

Manual wire-strippers have two blades that cross one, another like scissors. Each blade has a notch so that together they cut around the conductor in the middle. In addition, wire-strippers also have some means of adjustment so that different sizes of conductor can be stripped.

Before wire-strippers can be used, they must be set to the correct size. This can be done using a scrap or off-cut of wire of the correct size. With the jaws of the wire-strippers together, turn the adjustment screw until the hole in the jaws is just bigger than the conductor to be stripped. Test the setting on the off-cut of wire, by placing the wire-strippers over the wire and squeezing the handles to close the jaws. Then slightly release the jaws and try to slide the insulation off, using the wire-strippers.

If the wire-strippers are correctly set, the insulation will come off easily and there will be no damage to the copper conductor. If the aperture is set too small, the insulation will slide off easily but the conductor may be damaged. If it is set too large, the insulation will not come off easily. A simple adjustment of the adjusting screw will correct these problems. Now, with the correct setting, the wires can be stripped safely.

Other types of wire-stripper will have pre-set stripping holes so that selecting the correct hole will ensure that the depth of cut is correct every time.

Once the conductors are stripped they are ready to be connected to the electrical equipment.

Cable-strippers with pre-set holes

Terminating PVC/PVC flat profile cable

Tools required

Dependent on method used:

- electrician's knife
- side-cutters
- pliers.

Pliers, side-cutters, electrician's knife (from left to right)

Safety considerations

- Cuts to hands from use of knife
- Injury from slipping with cutters or pliers

The use of gloves and eye protection is recommended.

Method

Before the conductors can be connected, the outer sheath of the cable must be removed.

First, identify how much of the sheath should be removed from the cable. The purpose of the sheath is to provide some mechanical protection for the insulation on the conductors. Too much sheath, however, takes up space within the accessory and will put excess strain on the conductors. The sheath should, therefore, be stripped back almost to where the cable enters the accessory, leaving only 10–15 mm, a thumb's width, of sheath within the accessory.

There is more than one way to remove the outer sheath. Each has its advantages and disadvantages but, whichever method is used, care must be taken to avoid damage occurring to either the conductors or the insulation around the conductors.

Determining the point to which to strip the sheath

Method 1

Having decided on the length of sheath required, and with the cable in place, score the sheath at the point to which it is to be removed. This can be performed with an electrician's knife. Care should be taken not to cut into the cable.

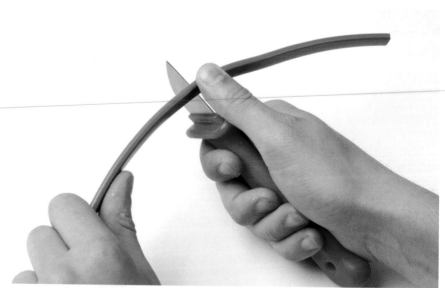

Scoring the outer sheath

From the end of the cable, snip down the centre of the cable, using a pair of side-cutters. Split the line to one side and the neutral to the other.

Determining the length to strip

4 Strip the copper sheath

The next step is to strip the copper sheath. Whilst there is more than one method of doing this, including using side-cutters, the easiest method is to use a specialist tool designed purely for this purpose.

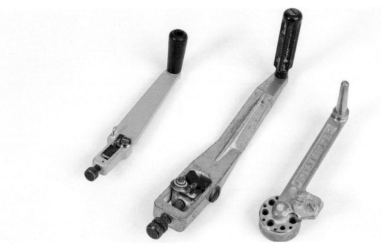

Different types of rotary strippers

Each of the tools works on the same principle. A small blade cuts into the metal sheath and peels off a small amount of copper sheath with each rotation.

JOI stripper in use

When the correct length of cable is reached, a pair of pliers is used to grip the cable just ahead of the stripper. This stops the stripper moving further down the cable.

Using pliers to end the stripping process

With the pliers in this position the stripper is turned again. As the stripper cannot move down the cable, it cuts a clean square edge to the cable. The sheath is pulled of and the cable is tapped gently to remove any powdered insulation that may be attached to the conductors.

5 Fit the gland to the cable
Push the gland onto the cable but do not tighten it.

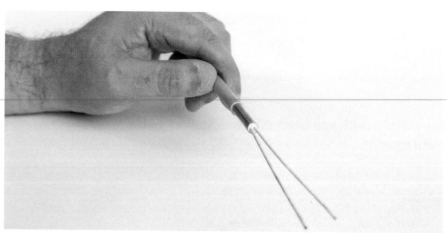
Cable ready for fitting pot

6 Fit the sealing pot to the cable
The sealing pot of the correct size is screwed onto the cable. A pot wrench is specifically designed for this job.

Using the type of pot wrench shown on the left requires that the wrench is tightened against the gland body, with the pot between the gland and the pot wrench.

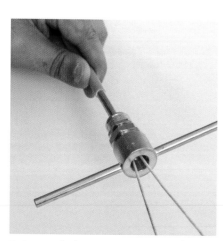
Pot wrench that uses the gland

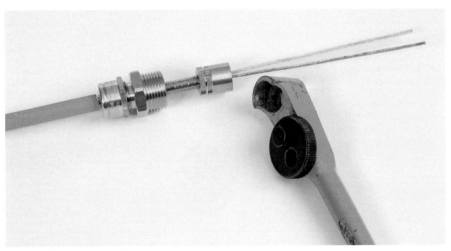

One-handed pot wrench

This type of pot wrench does not require a gland to be fitted to the pot. The wheel on the pot wrench locks against the knurled edge of the pot, thus locking the pot to the wrench.

With either type, pressure needs to be applied to get the pot started. As the pot is turned, it cuts a thread on the cable sheath.

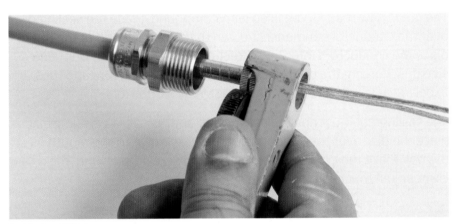

Pot being fitted

Once the cable end gets to the base of the pot, stop turning.

Inside the pot

If an earth tail pot is used, the earth tail must be aligned in the correct position, relative to the conductors.

Alignment of earth-tailed pot

Once the pot is in the correct position, remove the pot wrench. Turn the cable so that the open end is facing downwards, then tap the pot to ensure that any loose material in the sealing pot falls out. Carefully inspect the inside of the sealing pot for any signs of debris.

The pot is now ready to be filled with compound. Place the sealing disc over the conductors and push it down to the bottom end of the pot to place the conductors into the correct position. Fill the pot from one side only, forcing the compound between the conductors and thus expelling any air in the pot and avoiding any air pockets.

The sealing pot must now be sealed with another plastic sealing disc. Place the disc over the conductors and press down to the mouth of the pot. Place the crimper over the pot and screw down until the crimper just touches the pot.

Crimper in place ready to crimp

With a pair of pliers, pull gently on each conductor, in turn, to straighten it.

BW gland

These glands are for indoor use as they have no seal. The BW glands are available in two- and three-part configurations.

Two-part BW gland

The three-part gland has a cone ring as the extra component. The cone ring is tapered so that, as the gland is tightened, the cone ring is forced down onto the cone, tightening against the armour.

CW gland

These glands can be used outside, A CW gland has a seal, which seals the gland to the cable sheath. CW glands are available in three- and four-part configurations; again, the difference is the cone ring.

Four-part CW gland

Whilst the gland may be terminated onto the cable before any connection to the equipment is made, there are occasions when armoured cable has to be installed between fixed points without any tolerance or slack. Larger cables are more difficult to handle, so careful marking out of the cable is important. The following method assumes that the cable is be fitted to electrical equipment that has already been fixed to a wall.

These are the steps in the process of terminating an armoured cable.

1 Prepare the cable.
2 Fit the shroud.
3 Fit the lower part of the gland to the electrical equipment.
4 Ensure the gland is earthed as this will in turn earth the armour.
5 Cut the armour to the correct length.
6 Fit the upper part of the gland to the cable.
7 Mark up and remove the outer sheath.
8 Spread the armour strands.
9 Fit the gland parts together.
10 Remove the inner sleeve.
11 Terminate the cores to the electrical equipment.

1 Prepare the cable

The cable is cut to length, allowing enough at the end to terminate into the electrical equipment. Armoured cables are cut with either a hacksaw or with ratchet cutters designed to cut armoured cable.

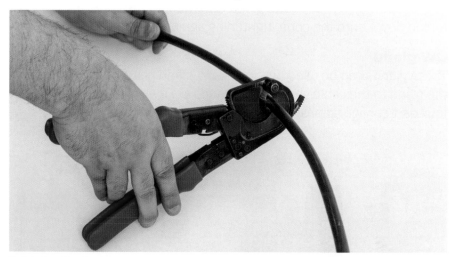

Ratchet cutter in use

2 Fit the shroud

Whilst the shroud can be fitted any time before step 8, it is best to cut the shroud top to the correct size before the outer sheath is removed. See the method of fitting the shroud, as described on page 224.

3 and 4 Fit the lower part of the gland to the electrical equipment and earth the armour

Steps 3 and 4 are integral in this method of termination. In other methods, where the gland is fitted to the cable before fitting to the enclosure, they are separate steps.

To earth the armour, the brass gland body (lower part) is passed through an earth tag and then into the accessory. It is secured in place with a lock ring or lock nut, using a pipe wrench. Before this can be done, a hole must be drilled into the accessory so that a bolt can be placed through the earth tag and the accessory. The earth 'fly-lead' can then be secured inside the accessory. This enables a connection to the installation earthing to be made.

It is important to ensure that the paint is removed from the accessory at the point of connection between the earth tag and the casing. Also, make sure that the bolt is installed with the thread and nut on the inside of the accessory, as this will make the connection tamper-proof.

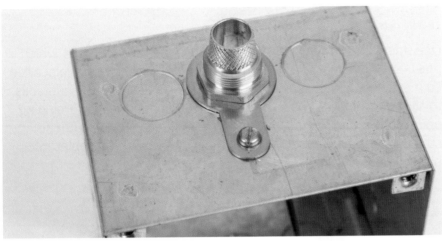

Gland body correctly fitted to steel enclosure.

The 'fly-lead' is made up of a suitably sized conductor, fitted with a ring-type crimp connector. This enables the securing bolt to act like a pillar terminal – the crimped end of the fly-lead is connected here, while the other end is terminated as a normal protective conductor. With the gland in place, the shroud can be brought up to cover the gland. There should be no strands of armour visible with the shroud in place.

ACTIVITY

Why is it necessary to remove any paint around the hole before terminating the cable?

Armour gland with fly lead

An alternative to using the earth tag is to fit a special nut called a Piranha nut. This replaces the lock ring or lock nut and removes the need for drilling extra holes in the accessory. It also provides a means of connecting an earth lead to the gland.

Piranha nuts

One side of the Piranha nut has small, sharp edges, like teeth, which bite into the enclosure as the nut is tightened, giving a good electrical connection. The nut also has threaded holes on each of the nut faces, to allow machine screws to be secured to the nut, thereby providing points for connecting the earthing fly-lead directly to the nut. This method is especially useful when terminating armoured cables into plastic boxes.

SWA cable terminated into plastic box   Purpose designed earth link plate

5 Cut the armour to the correct length

Before step 5 can be completed, the cable needs to be marked at the correct position. For this, the cable is held against the gland body that has been fitted to the electrical enclosure.

A marking is made which is approximately the thickness of one of the armour wires above the base of the cone of the gland.

Cable held against gland showing mark

Using a hacksaw, cut all the way around the cable, cutting through the outer sheath; continue cutting until each of the armour wires is cut to a depth of approximately half. Be careful not to cut too deep and thus damage the insulation of the conductors, but make sure that all of the armour wires are cut or knicked.

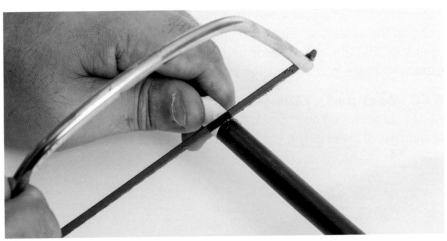

Cutting the armour

This step may also be accomplished by using an armoured cable-stripper, which looks rather like a pipe-cutter but has a blade to cut the outer sheath and score the armour wires.

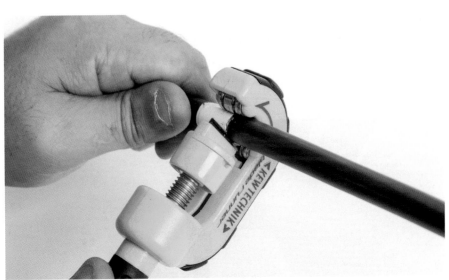

Armoured cable-stripper in use

Remove the outer sheath, down to the cut point, by using a stripping knife. Hold the cable end in one hand, hold the knife across the cable at a slight angle and cut the outer sheath, pushing away from you.

Cutting away the outer sheath

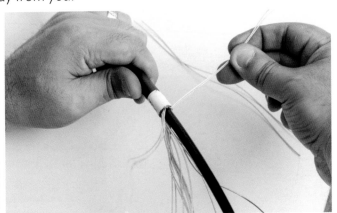

Breaking the armours off

Now remove the armour wires from the cable. This is achieved by bending the wires back and forth until they snap. Care must be taken to ensure that the armour ends under the sheath are not distorted, otherwise there may be problems when fitting the gland together.

6 Fit upper part of the gland to the cable

At this stage the shroud should have been fitted to the cable. The upper part or parts of the gland should now be slid over the cable.

Shroud and gland body on cable

7 Mark up and remove the outer sheath

Once again the cable is offered up against the gland body, now fitted to the electrical enclosure, and a mark is made just above the top of the cone.

How to mark up the sheath

At the mark, the outer sheath is cut right round the cable, using a knife, a hacksaw or a SWA stripper. The small piece of sheath is then removed.

Cable with sheath removed

It is recommended that Cat 6 patch leads are purchased, due to the fact that hand-making a Cat 6 patch lead is difficult and the result is likely to be of a poor quality.

TIA Wiring standards

T568A standard

RJ45 Pin	Wire colour (T568A)	Wire diagram (T568A)
1	White/Green	
2	Green	
3	White/Orange	
4	Blue	
5	White/Blue	
6	Orange	
7	White/Brown	
8	Brown	

T568B standard

RJ45 Pin	Wire colour (T568B)	Wire diagram (T568B)
1	White/Orange	
2	Orange	
3	White/Green	
4	Blue	
5	White/Blue	
6	Green	
7	White/Brown	
8	Brown	

Pins on an RJ45 plug are numbered from left to right, when the plug is held with the latch facing downwards and the pins on top of the plug.

RJ45 plug

SmartScreen Unit 306
Handout 18

Fibre-optic cable

Fibre-optic cables are used in data and telecoms applications. The most common connector types are:

ST – straight tip

SC – subscriber connector.

LC – Lucent connector, a small form factor SC type connector with a locking mechanism developed by Lucent

FC – fibre connector, less common these days as prone to vibration loosening

SC, LC and ST type connectors

Whatever type of connector is used, the termination process is similar. It is important that all dust is excluded from the termination. The environment when terminating fibre optics must be scrupulously clean. As the process of termination varies slightly from manufacturer to manufacturer, it is important to refer to manufacturer's instructions before terminating the fibre optic.

Tools required:

- crimp tool
- die sets to suit type of connector
- stripper tools
- kevlar shears
- cable holder
- cleave tool.

3 Fitting the termination

STEP 1 Open the cable clamp of the cable-holder assembly, and position the fibre (with the cleaved end facing the connector) inside the clamp. Move the fibre so that the end of the fibre is level with the front of the arm of the cable-holder assembly and, holding the fibre in place, close the clamp.

STEP 2 Carefully insert the fibre into the plunger of the connector assembly until the fibre reaches the internal fibre. Make sure that the remaining mark on the fibre enters the plunger.

STEP 3 The resultant bend in the fibre should hold the inserted fibre against the internal fibre. The fibre coating must enter the small tube that was installed in Step 4 of the 'Preparation and stripping' section. Make sure that the start of the fibre coating is not caught on the entry of the small tubing.

STEP 4 Squeeze the handles of the hand tool until the ratchet releases. Allow the handles to open fully.

With the connector assembly in the cable-holder assembly, position the ferrule or termination cover in the upper cavity of the front die and the plunger in the upper cavity of the rear die.

Gently push the fibre towards the connector assembly to make sure that the fibre is still touching the bottom and, then, slowly squeeze the tool handles together until the ratchet releases. Allow the handles to open fully and remove the connector from the dies.

Position the plunger of the connector assembly in the first (smallest) cavity of the front die, with the knurl against the edge of the groove in the die, and the ferrule or termination cover pointing in the direction of the arrow.

Slowly squeeze the tool handles together until the ratchet releases. Allow the handles to open fully and remove the connector assembly from the die.

STEP 5 Slide the bare buffer boot over the plunger until the boot butts up against the connector assembly. Remove connector assembly from the cable-holder assembly.

STEP 6 Align the key of the connector housing with the chamfered edges of the connector assembly. Slide the housing over the assembly until it snaps in place. DO NOT force the components.

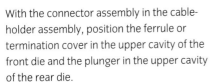

KEY POINT

The arrows marked on the front die indicate the direction in which the ferrule or termination cover must be pointing when the connector is positioned in that cavity. For proper placement, and to avoid damage to the fibre, observe the direction of the arrows.

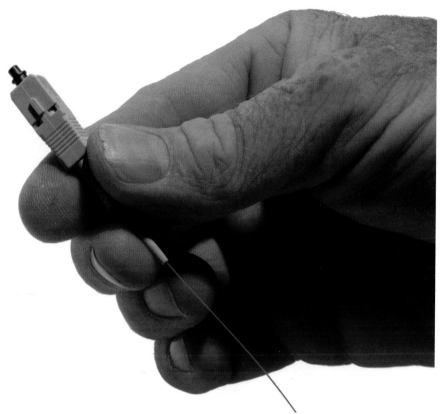

STEP 7 The fibre is now terminated, ready to be tested.

After the termination is complete, the ferrule end-face must be inspected for cleanliness, using a 200 × microscope. See the paragraph at the beginning of this section about the dangers of looking down a fibre-optic cable.

It can be seen from the foregoing that the process of terminating a fibre-optic cable is complex, but is a skill well worth obtaining. It is only by following the process and being scrupulously clean, that acceptable terminations can be made.

GLOSSARY

B

Bend radius This is a measurement of the tightness of the curve that can be applied when bending a cable. If the internal radius is too small then the cable will be subjected to undue stress and strain and may become damaged during use.

Bit The part of the drill that does the cutting.

Broadknife A tool rather like a wallpaper scraper but much more pliable.

Busbars Busbars are normally made from copper and are solid bars, used in place of wiring, to carry larger amounts of current.

C

Chuck The part of the drill used for holding the drill bit.

Closure A switch is typically referred to as closed when it is switched on and open when switched off.

CompEx Scheme The CompEx Scheme is now the global solution for validating core competency of employees and contract staff of major users in the gas, oil and chemical sectors. This covers both offshore and onshore activities.

Containment system The use of conduit and trunking when wiring is installed, to provide a level of mechanical protection to cables.

Cores The term 'cores' is used to identify the number of insulated conductors within the wire or cable.

cpc This is a circuit protective conductor. This is the conductor that performs part of the function of protective earthing, to keep persons safe under fault conditions.

Cross-talk Cross-talk is a bleeding of signal from one conductor to another, through electromagnetic induction. Twisting pairs of cable reduces this cross-talk.

D

Datum line A reference point or line from which multiple measurements are made.

Drop The amount of paper that is required to cover a strip from the ceiling to the floor.

Duty holder A person with legal responsibilities and who is also responsible for managing risk.

E

Eddy current An eddy current is a circular current that is caused by the electromagnetic field of the conductor inducing an emf. These currents circulate, first one way and then the other, and can result in a lot of heat being generated.

Egress An alternative term for exiting or leaving but is commonly used in technical publications.

Electrical buyer The person who places the purchase orders with suppliers for electrical equipment. They are often required to source and negotiate deals to secure the best price and delivery for a wide selection of electrical accessories and components. Big discounts can be given by the supplier if the electrical buyer can place large orders.

emf Electromotive force is the driving force of electrical energy and it is measured in volts.

EMI Electromagnetic Interference (EMI) is caused by some a.c. circuits, due to the current flow and resulting electromagnetic effect. EMI can cause surges and spikes, which can have an impact on circuits where the conductors are run alongside other circuits.

Enabling Act An enabling Act allows the Secretary of State to make further laws (regulations) without the need to pass another Act of Parliament.

Extraneous conductive part A conductive part that is not part of the installation but is connected to earth and under fault conditions may be at a different potential to the rest of the metal parts within an installation.

F

Ferromagnetic A ferromagnetic material is one that can be turned into a magnet. Typical examples include iron and steel, both of which are used to make electrical enclosures.

Floor area served This takes into account the size of the room where the socket outlet is located. In large open-plan areas, the length of the appliance flex needs to be taken into account as per regulation 553.1.7 of BS 7671.

Halogen A group of elements in the periodic table which are classed as toxic.

I

Installer Anyone who performs electrical installation work, irrespective of job title.

J

Just in time A large-quantity order is placed with a supplier, with agreement on a staged delivery. This means that the total order is placed at the start, to take advantage of quantity discounts, but the supplier stores the materials and only delivers the required quantity at the agreed times throughout the project.

L

Live BS 7671 defines a live part as a conductor or conductive part that is intended to be energised in normal use. This means that a line conductor and a neutral conductor can be both referred to a live conductors. Any conductor that is providing a protective function is not a live part.

Live conductor A conductor intended to be energised in normal service, and therefore includes a neutral conductor.

M

Marking out Making measurements and marking positions on the walls and ceilings, to ensure that equipment and accessories are installed in the correct places.

Method statement This is a written statement that identifies how the work activity will be performed and what safety measures will be employed to control the risks identified in the risk assessment, as well as detailing tools, equipment and personnel required. It may also detail other considerations such as access requirements, time frames and potential impacts of the work.

N

Noggin A piece of wood that is used to bridge the distance between studs or joists by being fitted in between the studs or joists. Noggins can be used for securing accessories in place.

O

On-load switching This requires that the switch can make and break its full current rating. Not all switches can break their full load rating.

Overcurrent BS 7671 defines an overcurrent as a current exceeding the rated current.

P

Pattress The recessed container behind an electrical fitting, such as a socket; often referred to as a back box.

Person ordering the work This may be the client, an architect, a consultant or a main contractor, depending on the job organisation.

Personal protective equipment (PPE) Protective clothing, helmets, goggles, or other garments or equipment designed to protect the wearer's body from injury.

Power tool Any tool that requires electrical energy to make it work. The electrical energy can be obtained from batteries or from a mains supply.

S

Safe system of work This is the integration of personnel, articles and substances in a pre-determined and considered method of working. A safe system of work takes proper account of the risks to employees and others who may be affected, such as visitors and contractors, and provides a formal framework to ensure that all of the steps necessary for safe working have been anticipated and implemented.

Swarf Chips and spirals of waste metal produced when cutting metal.

Sheath The sheath of a cable is the outer covering which not only holds the cable together, but also provides a basic level of mechanical protection against light damage that may be caused during installation or use.

Short shipments These occur when the supplier delivers an incomplete order. There are normally items on the original order still to be delivered, or the quantity is short and the balance is yet to come.

Skilled or instructed person (electrically) Skilled – A person with relevant education and experience to enable them to perceive risks and avoid hazards which electricity can create. Instructed – A person adequately advised or supervised by electrically skilled persons to enable them to perceive risks and avoid dangers which electricity can create.

Solder creep Sometimes called 'cold flow', 'solder creep' can occur when the termination is under constant mechanical stress and the solder can literally move or 'creep'. Incidence of creep increases with temperature.

Statue law Law that has been laid down by Parliament as Acts of Parliament.

Stud This is the support for a partition wall and normally runs from floor to ceiling. They are sometimes joined, by cross-members called noggins, to provide additional strength to the wall and provide mounting positions for accessories.

U

UPS An uninterruptable power supply is often used where computer systems use a central server. The UPS provides sufficient power to the server to enable a controlled shut-down of the server in the event of a mains failure.

V

VDE An acronym of Verband der Elektrotechnik (originally the Association of German Electrical Engineers and now the Association for Electrical, Electronic and Information Technologies, in Germany) which is responsible for testing and certifying tools and appliances.

W

Way When used in relationship to distribution boards, the way is the maximum number of single-phase circuits that can be connected to the board.

ANSWERS TO ACTIVITIES AND KNOWLEDGE CHECKS

Answers to activities and knowledge checks are given below. Where answers are not given it is because they reflect individual learner responses.

UNIT 305 PRACTICES AND PROCEDURES FOR THE PREPARATION AND INSTALLATION OF WIRING SYSTEMS AND ELECTROTECHNICAL EQUIPMENT

Answers to margin activities

Page

2   NVQ level 3.

4   No, only incidents that happen in the workplace.

7   BS 7671.

8   The first aid kit should contain:

- a leaflet giving general guidance on first aid (e.g. HSE's leaflet Basic advice on first aid at work)
- 20 individually wrapped sterile plasters (of assorted sizes), appropriate to the type of work (you can provide hypoallergenic plasters if necessary)
- two sterile eye pads
- four individually wrapped triangular bandages, preferably sterile
- six safety pins
- two large, individually wrapped, sterile, unmedicated wound dressings
- six medium-sized, individually wrapped, sterile, unmedicated wound dressings
- at least three pairs of disposable gloves

9   a) It is height above the floor that matters.

11   Fencing, security guards, control of access to the site, lighting, etc.

14   Place receptacles for the collection of waste both in and near to canteen areas. Make sure these are cleared daily.

18   14.8%

Page

21   a)   Secure lockup as part of site offices.

   b)   Secure compound kept locked when not supervised.

27   Store in a shed or keep in van.

30   Carry out safe isolation procedure. Lock off the circuit breaker supply or remove fuse/disconnect circuit. Secure the distribution board. Place a warning notice on board.

32   No, these meters have a maximum operating voltage of 24 V. GS 38 only applies above 50 V a.c or 120 V d.c.

34   b)   switch disconnector.

   d)   plug and socket.

37   A suitable device is the 'Castell' lock.

38   Six:

L1-L2

L1-L3

L2-L3

and each line to earth.

44   NYY-J cable (armoured, sunlight resistant).

44   As a means of identification.

46   The metal sheath is likely to corrode due to acid attack or electrolytic action.

48   Other fire resistant types include Firetuf.

50   Numbers.

53   Yes, they can.

55   IDC (insulation displacement connectors).

57   Decorative lighting.

58 They can be installed In ducts.

61 The flexible conduit cannot be relied upon for earth continuity.

65 Adaptable boxes.

67 As the back plate is fixed first, two hands are available to hold the conduit and tighten the screws.

69 It has a limited capacity and when joined to larger sizes, reducers are required. Most metal accessory boxes only have 20 mm knockouts. The reduced demand for 16 mm means that it is expensive compared to 20 mm.

71 75 × 50 (factor 1555) and 100 × 38 (factor 1542)

75 Cost. An electrician cannot make one for the same cost as a bought one.

Finish. A home-made one never looks as good as a factory produced unit.

77 A fillet.

79 Busbars allow easy repositioning of machines without major rewiring.

81 The surrounding air temperature.

83 4 = prevention of entry of 1 mm wire.

X = no protection against moisture.

86 3 × the switch wires, line, neutral and earth.

88 Nothing really, consumer units are generally used in domestic installations and distribution boards in commercial or industrial installations, but a consumer unit is just a type of distribution board.

90 A properly set up laser level will give an exact level all around the room. The use of a small boat level will introduce errors.

92 Forward.

93 Reamer or possibly a cone cutter.

98 Internal spring.

100 Lump hammer.

101 Various answers showing the effects of improper use:

a) drill bit will burn out

b) drill will not cut successfully as point is more or less blunt. Drill will wander over surface

c) drill will grab as it breaks through and create a deformed hole

d) drill will jam or burn wood.

102 1) plain drilling.

2) hammer drilling.

3) screwdriving.

105 Flat bit for cutting wood.

106 Over-tightening and snapping terminal.

109 A non-setting material.

111 Round head, to prevent raised edges of countersunk screws damaging cables.

112 Coach bolt.

115 The material goes off after time and becomes less effective.

119 A wallpaper-pasting table, usually a cheap construction of batten and hardboard, that folds up.

126 A fault of negligible impedance occurring between live conductors of the same circuit.

128 D.C. This is then converted to a.c. by an inverter.

132 No need to drill them to attach lugs.

134 Iron losses.

136 Water, humidity, high ambient temperature, movement.

138 Screws will identify the cut board as an access trap. Frequent re-nailing may split the board.

143 Use high-temperature flex or sheathing.

146 In series.

151 Mains-fed shaver (no earth on secondary side).

164 Damage to the system when making holes in the ceiling for various purposes.

Outcome knowledge check answers — short-answer questions

Page 172

1   Employers, employees and self-employed

2   Health and Safety (First Aid) Regulations 1981

3   The construction of a building will have an effect on the choice of a wiring system and installation methods used. Starting work and finding that a floor is concrete instead of wood will have a considerable effect on the materials required and time taken.

4   Proceed immediately to the assembly area.

5   A light switch is a micro gap switch with insufficient contact separation for Isolation. In addition, it cannot be locked in the OFF position and the ON and OFF position are not clearly and reliably indicated.

Page 172

6   Remove and keep fuses on person, or disconnect conductors from terminals.

7   There is always a chance that there is a potential (voltage) between neutral and earth.

8   By testing on a known supply or by using a proving unit.

9   Other alternatives are acceptable.

a) dado trunking, electrack

b) overhead busbar trunking

c) flat pvc/pvc

d) MICC cable.

10 Magnesium oxide insulation and copper conductors and sheath.

Outcome knowledge check answers – multiple-choice quiz questions for learning outcomes 1–6

Page 173–174

1 c)	2 c)	3 d)	4 b)	5 d)
6 d)	7 c)	8 b)	9 c)	10 b)
11 d)	12 d)	13 b)	14 d)	15 b)

Outcome knowledge check answers – multiple-choice questions

Page 175–176

1 c)	2 d)	3 c)	4 b)	5 a)
6 b)	7 b)	8 b)	9 c)	10 c)
11 b)	12 d)			

UNIT 306 PRINCIPLES, PRACTICES AND LEGISLATION FOR THE TERMINATION AND CONNECTION OF CONDUCTORS, CABLES AND CORDS IN ELECTRICAL SYSTEMS

Activity answers

Page

179 A voltage indicator complying with GS38 and a proving unit.

182 Test instruments are used to identify if electrical systems are safe and therefore it is important that all test instruments are checked regularly for accuracy. This regular check is part of the calibration process.

187 Systems with battery backups such as fire alarm systems. Systems with uninterruptable power supplies (UPS).

190 The EAWR is a statutory document and the IET wiring regulations gives a practical interpretation of the EAWR regulations.

192 The disconnected end would be at mains potential and if touched would give a severe electric shock.

193 It provides a defence if all practical efforts have been made to prevent an incident.

198 Site meeting to agree changes with parties involved and agree variation to original costs. Obtain variation order.

199 Despite most people's opinion it does not rain all the time in the UK and it should be possible to carry out the work at a dry time. However the use of suitable tarpaulins and tents should provide adequate coverage.

203 One method would be 3-plate wiring with connections at the ceiling rose or batten holders.

Other systems have connections within the pattress.

205 No.

208 The equipment has no hole greater than 1mm and is protected against splashes of water from all directions.

209 A junior hacksaw. Don't just break the whole section out.

Page

210 As the socket plate is pushed back there is a tendency for the connection to loosen, causing a high resistance joint. In addition if the conductors have to be untwisted for testing purposes this process weakens the conductor.

214 Insufficient crimped area.

Damage (cut) to crimp and conductor.

217 The damaged insulation should be removed and replaced with new insulation of an equivalent value to the original.

219 Amalgamating tape. This sounds obvious but many tapes used as insulation tapes are actually coloured identification tapes.

220 Dust mask.

223 A torque wrench. It is used to tighten terminals to a pre-set specific value.

225 A suitable coupler would be a waterproof flex connector.

227 A common fire-resistant cable would be Firetuf.

233 1.5 mm^2 – MICC cable has a higher current rating than PVC.

240 If both ends are made off, there may be a fault on one pot and it is impossible to tell which one it is. Damp can be detected by using an IR tester and applying heat to the pot. If moisture is present the IR value will fall.

241 The sheath becomes work hardened and is difficult to reset.

245 The paint can act as an insulator between the gland and the switchgear.

250 Training is available from network distribution companies such as Scottish and Southern.

261 The information is available from wholesalers, websites or wholesalers, catalogues.

Knowledge check answers — short-answer questions

Pages 273–274

1

Number	Document	Statutory or non-statutory (S/NS)
1	Electricity at Work Regulations 1989	S
2	IET Wiring Regulations BS 7671 2008	N
3.	IET On-Site Guide	N
4	Health and Safety at Work Act 1974	S
5	COSHH Regulations	S
6	IET Guidance Note 3	N

2 A variation order

3 Electrically continuous and mechanically sound.

4 Replace PVC insulation with high-temperature insulation or use high-temperature cable for final connection.

5 Overheating and burning out the cable or terminal

6 Aluminium is relatively soft compared to copper and small cables may be cut through if overtightened

7 1. No copper showing

2. Terminal screw on copper, not insulation

3. Terminal as full as possible

4. Screw tight but not overtight.

8 a) in any order

b) The N & cpc must be in the same order as the Line.

9 Loop or loop in.

10 An earth washer, commonly known as a banjo is bolted to the box.

11 Strain pillars

12 Protection not considered, so IP4X gives protection against 1mm wire, but moisture protection not considered.

13 Nothing mandatory, but about 10–15mm

14 They are much quicker to fix.

15 The flex may be pushed out as the stud nut is tightened up.

INDEX